Digital Food Activism

This book explores the role of digital media technologies in creating new forms of consumer activism and engagement with food, eating and food systems. Food is an increasingly prominent subject of engagement online, from the aesthetics of cooking to the ethics of shopping.

This book adopts a multi-disciplinary approach, bringing together food studies, and science and technology studies. The role of social media, apps, and other online technologies is considered in relation to activist and consumer issues in the UK, Australia, Europe and South America. *Digital Food Activism* explores a variety of contemporary topics, including Twitter and diabetes, hashtag activism and the prospect of 3D printed food.

Tanja Schneider is a Senior Lecturer in Sociology at the Institute of Sociology (SfS-HSG), University of St. Gallen, Switzerland, and a Research Associate, Institute for Science, Innovation and Society and Institute of Social and Cultural Anthropology, School of Anthropology and Museum Ethnography, University of Oxford, UK.

Karin Eli is a Postdoctoral Research Fellow at the Institute of Social and Cultural Anthropology, School of Anthropology and Museum Ethnography, University of Oxford, UK.

Catherine Dolan is a Reader in Anthropology in the Department of Anthropology and Sociology, SOAS, University of London, UK.

Stanley Ulijaszek is Professor of Human Ecology and Director of the Unit for Biocultural Variation and Obesity, Institute of Social and Cultural Anthropology, School of Anthropology and Museum Ethnography, University of Oxford, UK.

Critical Food Studies

Series editor: Michael K. Goodman

University of Reading, UK

The study of food has seldom been more pressing or prescient. From the intensifying globalization of food, a world-wide food crisis and the continuing inequalities of its production and consumption, to food's exploding media presence, and its growing re-connections to places and people through 'alternative food movements', this series promotes critical explorations of contemporary food cultures and politics. Building on previous but disparate scholarship, its overall aims are to develop innovative and theoretical lenses and empirical material in order to contribute to – but also begin to more fully delineate – the confines and confluences of an agenda of critical food research and writing.

Of particular concern are original theoretical and empirical treatments of the materializations of food politics, meanings and representations, the shifting political economies and ecologies of food production and consumption and the growing transgressions between alternative and corporatist food networks.

For a full list of titles in this series, please visit www.routledge.com/Critical-Food-Studies/book-series/CFS

Practising Empowerment in Post-Apartheid South Africa
Wine, Ethics and Development
Agatha Herman

Digital Food Activism
Edited by Tanja Schneider, Karin Eli, Catherine Dolan and Stanley Ulijaszek

Forthcoming

Children, Nature and Food
Organising Eating in School
Mara Miele and Monica Truninger

Hunger and Postcolonial Writing
Muzna Rahman

Taste, Waste and the New Materiality of Food
Bethaney Turner

Digital Food Activism

Edited by
Tanja Schneider, Karin Eli,
Catherine Dolan and
Stanley Ulijaszek

Routledge
Taylor & Francis Group

LONDON AND NEW YORK

First published 2018
by Routledge

2 Park Square, Milton Park, Abingdon, Oxfordshire OX14 4RN
52 Vanderbilt Avenue, New York, NY 10017

Routledge is an imprint of the Taylor & Francis Group, an informa business

First issued in paperback 2019

British Library Cataloguing in Publication Data
A catalogue record for this book is available from the British Library

Library of Congress Cataloging in Publication Data
A catalog record for this book has been requested

ISBN: 978-1-138-08832-0 (hbk)
ISBN: 978-0-367-88881-7 (pbk)

Typeset in Times New Roman
by Wearset Ltd, Boldon, Tyne and Wear

Contents

Illustrations

Figures

Tables

Contributors

Mariano Beguerisse-Díaz is a Senior Research Fellow in the Mathematical Institute at the University of Oxford, UK. His primary research interest is developing mathematical methods for the tractable representation of complex data systems. In particular, he is interested in directed networks, network inference, the dynamics on/of networks, temporal networks, and going from meta-data to full-data descriptions. He develops techniques for analysing data and models using dynamical systems, matrix analysis, probability, topological data analysis, numerical analysis, statistics and information retrieval to work on problems from biology, social science, economics, and engineering.

Melissa L. Caldwell is Professor of Anthropology at the University of California, Santa Cruz, USA, and Editor of *Gastronomica: The Journal of Critical Food Studies*. Her research and teaching are situated at the nexus of food studies, political economy, poverty and welfare, and social justice. Her long-term ethnographic research in Russia examines the entanglement of political systems in the most ordinary spaces and dimensions of people's lives, and in turn, how individuals represent, interpret, and experience their relationships with the state and their fellow citizens. Her current research focuses on hacking and creative economies as modes of social activism. She is the author of *Not by Bread Alone: Social Support in the New Russia* (University of California Press, 2004), *Dacha Idylls: Living Organically in Russia's Countryside* (University of California Press, 2011), and *Living Faithfully in an Unjust World: Compassionate Care in Russia* (University of California Press, 2016); and is the editor of several volumes on food.

Catherine Dolan is on the faculty of Anthropology at SOAS, University of London, UK, and holds fellowships at Said Business School, Oxford, and Green Templeton College, University of Oxford, Oxford, UK. She is an anthropologist specializing in contemporary forms of moral capitalism, including Fair Trade, inclusive development, and corporate social responsibility, particularly in East Africa. Much of her research has focused on the cultural economies of alternative food provisioning, and how processes of labelling, certification, and retail consolidation have reshaped producer/ consumer relations. Her books include *The Anthropology of Corporate Social*

Responsibility (Berghahn Books, 2016, with D. Rajak) and *Ethical Sourcing in the Global Food System* (Earthscan, 2006, with S. Barrientos), and she has published widely in anthropology and development studies on agro-food commodity chains, sustainable agriculture, and alternative trade networks. She is a founding member of the Oxford Food Governance Group and the Centre for New Economies of Development (www.responsiblebop.com), a UK-based consortium that critically examines the implications of market-centred paradigms for development.

Karin Eli is a medical anthropologist based at the Institute of Social and Cultural Anthropology, University of Oxford, Oxford, UK. Her research focuses on eating disorders, obesity, food governance, social class and embodiment. Karin works at the intersections of the social sciences, the clinical sciences, and the humanities. Her publications include numerous articles in peer-reviewed journals, alongside chapters in edited volumes, including *Dietary Sugars and Health* (CRC Press, 2014) and *Careful Eating: Bodies, Food and Care* (Ashgate, 2015). Together with Stanley Ulijaszek, Karin co-edited the volume *Obesity, Eating Disorders and the Media* (Ashgate, 2014). She is the co-editor of the forthcoming special issue of *Transcultural Psychiatry* on 'Anthropological Perspectives on Eating Disorders'.

Ryan Alison Foley recently completed her DPhil (PhD) at the Institute of Social and Cultural Anthropology, University of Oxford, Oxford, UK, where she studied the role of values in Italian social cooperative practice. She has previously written an article and a number of book reviews for the *Journal of the Anthropological Society of Oxford Online*. Her primary research interests include economic ideologies, the construction of needs, and cooperative business practices.

Eva Giraud is Lecturer in Media, Communication and Culture at Keele University, Keele, UK. Her research explores the mediation of environmental and intersectional politics, with a particular focus on 'contentious' activism. Her work has been published in journals including *Theory, Culture & Society*, *Subjectivity, Convergence*, and *Feminist Review*.

Michael Goodman is Professor of Human Geography at the University of Reading, Reading, UK. His research focuses on alternative food networks, food geographies and the cultural politics of consumption, humanitarianism and food. He has written several books, including *Alternative Food Networks: Knowledge, Practice and Politics* (Routledge, 2011, with D. Goodman and M. DuPuis), *Food Transgressions: Making Sense of Contemporary Food Politics* (Ashgate, 2014, with C. Sage), *Consuming Space: Placing Consumption in Perspective* (Ashgate, 2010, with D. Goodman and M. Redclift), and *Food Geographies: An Introduction* (Bloomsbury, 2016, with M. Kneafsey, D. Maye and L. Holloway). He is also the series editor for Routledge's Critical Food Politics and Bloomsbury's Contemporary Food Studies series.

Tania Lewis is Professor in the School of Media and Communication at RMIT University, Melbourne, Australia. Her research critically engages with the politics of lifestyle, sustainability and consumption, and with global digital and media cultures. Tania has published over 50 journal articles and chapters and is the author of *Smart Living: Lifestyle Media and Popular Expertise* (Peter Lang, 2008), and co-author of *Telemodernities: Television and Transforming Lives in Asia* (Duke University Press, 2016, with F. Martin and W. Sun) and *Digital Ethnography: Principles and Practices* (Sage, 2015, with S. Pink, H. Horst, J. Postill, L. Hjorth and J. Tacchi). She is also the editor and co-editor of four collections with Routledge, including *Ethical Consumption: A Critical Introduction* and *Green Asia: Ecocultures, Sustainable Lifestyles and Ethical Consumption*. A member of the Digital Ethnography Research Centre at RMIT, Tania has conducted a wide range of research, from video ethnographic studies of backyard permaculture and household recycling to qualitative research with TV audiences and industry in South and East Asia. She is currently a Chief Investigator on the project 'Work-Life Ecologies: Lifestyle, Sustainability, Practices'.

Javier Lezaun is James Martin Lecturer in Science and Technology Governance at the Institute for Science, Innovation and Society, and Associate Professor in the School of Anthropology and Museum Ethnography, at the University of Oxford, Oxford, UK. His work focuses on the legal and social implications of advances in the life sciences, including biotechnology and pharmaceutical research. He has researched the controversies over the use of genetic engineering methods in agricultural and food production. He recently directed BioProperty, an ERC-funded research programme on the role of property rights in biomedical research. He has also been Co-PI of the project 'Emerging Forms of Food Consumer Behaviour and Food Governance' at the University of Oxford.

Deborah Lupton is Centenary Research Professor in the News and Media Research Centre, Faculty of Arts and Design, University of Canberra, Canberra, Australia. Her latest books are *Medicine as Culture*, 3rd edn (Sage, 2012), *Fat* (Routledge, 2013), *Risk*, 2nd edn (Routledge, 2013), *The Social Worlds of the Unborn* (Palgrave Macmillan, 2013), *The Unborn Human* (editor, Open Humanities Press, 2013), *Digital Sociology* (Routledge, 2015), and *The Quantified Self: A Sociology of Self-Tracking* (Polity, 2016). Her current research interests all involve aspects of digital sociology: big data cultures, self-tracking practices, digitised pregnancy and parenting, the digital surveillance of children, 3D printing technologies, digitised academia, and digital health technologies.

Sarah Lyon is Associate Professor of Anthropology at the University of Kentucky, Lexington, USA. She is the author of *Coffee and Community: Maya Farmers and Fair Trade Markets* (University Press of Colorado, 2011), the winner of the Society for Economic Anthropology's Book Prize, and the co-editor of *Fair*

Trade and Social Justice: Global Ethnographies (New York University Press, 2010). She is currently conducting on-going research on coffee, gender, economic development, and the impact of certification and quality standards in Mexico. She is a member of the Fair Trade Colleges and Universities National Steering Committee. At the University of Kentucky, Sarah teaches courses on economic anthropology, business anthropology, ethical consumption, and globalization. She is the editor-in-chief of *Human Organization.*

Alana Mann is Senior Lecturer in the Department of Media and Communications, Faculty of Arts and Social Sciences at the University of Sydney, Sydney, Australia. Her research affiliations include the Sydney Environment Institute (SEI) and the Charles Perkins Centre for cardiovascular health. She sits on the executive committee of the Australian Food Sovereignty Alliance (AFSA), an organization dedicated to creating a more ecologically sound and fairer food system for all Australians. Her book on international food sovereignty campaigns, *Global Activism in Food Politics: Power Shift*, was published by Palgrave Macmillan in 2014.

Amy McLennan is a Research Associate at the School of Anthropology, University of Oxford, Oxford, UK, and a Senior Analyst in the Australian Government. Amy's work is underpinned by the view that collaborative intersectoral and interdisciplinary approaches are key to understanding and addressing some of today's major global health issues. Her background reflects this view: her academic training in the sciences and social sciences, and professional experience in the academic, policy, and private sectors. Her academic research to date has broadly focused on social aspects of food, nutrition, and nutrition-related non-communicable diseases, including obesity and diabetes.

Tanja Schneider is a Senior Lecturer at the Institute of Sociology at the University of St. Gallen, St. Gallen, Switzerland, and Research Associate in the Institute for Science, Innovation and Society and the Institute of Social and Cultural Anthropology (ISCA), both at the University of Oxford, Oxford, UK. Her research is situated at the intersections of science and technology studies, economic sociology, and critical food studies. Her food-related publications focus on the governance of genetically modified and functional foods in Europe, the marketing of health foods and healthy eating practices in Australia, and the development of digital forms of food activism, and have appeared in *Science as Culture, Geoforum, Information, Communication and Society, Health Sociology Review*, and *Consumption, Markets and Culture*. She has also contributed chapters to *The Practice of the Meal: Food, Families and the Market Place* (Routledge, 2016) and *Careful Eating: Bodies, Food and Care* (Ashgate, 2015). Tanja's current research explores the uptake of venture capital in food and food-related settings at the forefront of developing novel materials, technologies, and business models affecting food production, distribution, and consumption.

Bethaney Turner is Assistant Professor in the Faculty of Arts and Design at the University of Canberra, Canberra, Australia. Her current research explores the variety and complexity of the relationships between people and the food they grow, buy, and consume. From local community gardens to global debates about food security, her research analyses the role that food plays in the formation of subjectivities, practices of meaning-making and under-standings of place.

Stanley Ulijaszek is Professor of Human Ecology and Director of the Unit for Biocultural Variation and Obesity, at the School of Anthropology, University of Oxford, Oxford, UK. He is Associate Editor of *Homo: Journal of Comparative Human Biology*. He has specialised in nutritional anthropology since his first academic position in Biological Anthropology at the University of Cambridge, and has taken more evolutionary and biocultural approaches to nutrition and nutritional health since his appointment at University of Oxford in 1999. His work on nutritional ecology and anthropology has involved fieldwork and research in Papua New Guinea, the Cook Islands, and South Asia. Among his books are *Nutritional Anthropology* (Smith-Gordon, 1993); *Human Energetics in Biological Anthropology* (Cambridge University Press, 1995); *Cambridge Encyclopaedia of Human Growth and Development* (Cambridge University Press, 1998, with F. E. Johnston and M. A. Preece); *Holistic Anthropology* (Berghahn Books, 2007, with D. Parkin); *Evolving Human Nutrition* (Cambridge University Press, 2012, with S. Elton and N. Mann); *When Culture Impacts Health: Global Lessons for Effective Health Research* (Academic Press, 2013, with C. Banwell and J. Dixon); and *Population in the Human Sciences* (Oxford University Press, 2015, with P. Kraeger, B. Winney and C. Capelli). He presently conducts multidisciplinary research into the political ecology of nutrition and obesity globally, using anthropological, life history, epidemiological, political, and economic historical frameworks.

Katharina Witterhold is Senior Research Associate in the Department for Social Sciences, University of Siegen, Germany, on the research project 'Consumer Protection and Consumer Socialization of Refugees'. Her research focuses on consumer sociology with special interest in the mediatization and politicization of consumption.

Preface

The politics of eating bits and bytes

Mike Goodman

Learning what and how to eat is now a thoroughly digital affair. Learning what can be done with and through food is *also* now utterly digital in the food activism that has quickly colonised cyberspace. But what does this collection of 1's and 0's in the programmed form of mediated food politics mean for relationships to food, to activism, to the digital? What about our relationships to each other and to the natural world around us, to food knowledge and foodie 'things'? What about our understandings of ourselves and our own shifting, sometimes stable, connections to food and its digitised cultural politics? Can these unique collections of bits and bytes really change not just who we are and how we mediate food and food activism but actually modulate the material 'worlds of food' (Morgan *et al.*, 2006) in meaningful and substantive ways?

Exploring, playing with, bending, twisting, rearranging and critiquing these questions is at the heart of *Digital Food Activism* and its empirically-rich and theoretically-precocious take on the 'ontological experiments' that frame the contemporary online world of food politics. Filtered through the multi-variate lenses of food hacking, humorous food resistances, subaltern digitalities, 3D fabricated foods, community organising and online permaculture assemblages, this volume is the first of its kind to work through the tensions, openings and boundaries created through the cultural politics of food activism in the digital age.

Digital food activism – and the multiple forms it takes, as this volume shows us – offers up what, in discussing food media more broadly, Christine Barnes (2017) refers to as 'moments of possibility'. The moments of possibility embedded in digital food activism work as 'knowledge transfer devices' designed to get audiences to think differently about food and change behaviours to create more diverse food systems. These digital spaces, in other cases, facilitate materially-significant networks to and from the 'real world' through activisms that include the human hands behind the keyboard and mouse, to the minds and bodies of those putting their hands in the dirt and more ethical food on their plate. Thus, online and offline resistances converge to co-create each other through fits and starts, through seamless connections, through the overt and the implicit, indeed, through the *multitudinous* ways that food's multitudes (Mol, 2008; Leer and Povlsen, 2016) are replicated each and every day in the 'connective action' of digital food activism.

Yet, this seeming democratisation of the foodscape is constrained by 'moments of *impossibility*', the antonym of Barnes' conceptual intervention (2017). Equal and opposite instances of 'not doing' or 'not being able to do' are also embedded within digital food activism. They work to constrain activists and eaters through the boundedness of what various techno-social platforms can and can't do, the inequalities of content creation and access, and the overarching imperatives of capitalist value generation that embed profit at the centre of digital platforms. Various forms of digital food activism are bounded by our ever-shrinking attention spans, the labour it takes to filter digital information from noise to signal and the growing difficulties of gaining the eyeballs needed to construct food activism. In these duelling moments of im/possibility, digital food activism offers up a 'meditated biopolitics' of food (Goodman *et al.*, 2017) that include within them sometimes fleeting, sometimes 'sticky', moments of contestation, hope and the possibilities of building 'other' food worlds. But because of how our digital universe is structured, digital food politics must also acquiesce to hegemonic algorithms, partake in Big Capitalism and even, presumably, shape the continuing echo-chamberisation of politics. These paradoxes and conflicts inherent in digital food activism – and the political, social and ecological stakes they raise – play out in various ways in the brilliant chapters that make up this volume.

A second set of tensions analysed here are those contained within the transparencies that some formats of digital food activism enable as part of their promise for better food worlds. Indeed, the transparency afforded by the digital's new ways of knowing food, nature, growers and other eater activists, entangles us in global food networks like never before. These connective actions deploy the digital as a force for 'good' in ways unimaginable by early food movements who had to work through handmade flyers at dusty health food stores. But of course, transparency in food assemblages is not all that new. For quite some time, scholars – some of whom are contributors to this volume (i.e. Dolan, Lyon) – have pointed out the ways that things like ethical food labels work to show us the story 'behind the food' in bids to facilitate more moral food economies. What is new then about digital food activism's bid for transparency? Surely, if anything, it is about the ease of shareability, the replicability and transferability of this information, its mobility, its possibilities of rich, thick description and imagery, and its almost instant ability to create (relatively free) global networks in the media ecologies of the digital.

However, even with digital food activism's offer of a techno-social transparency surrounding the food we eat, or *should* be eating, it is possible to miss critical concerns right in front of our eyes. For example, last week I ate a banana – brought to me by my friendly, multi-national food corporation – that was tagged as originating from farm '55116' and that urged me to digitally 'visit' the farm from whence it came. I took the bait and entered the farm number on a webpage – entitled 'Dole Earth' for obvious reasons – and came face to face with Columbian banana farm 55116. I was then treated to an interactive farm tour of all aspects of the banana harvest and, not surprisingly, what Dole is doing to protect workers and the planet. Andrés Jimenezu, one of the farm workers featured in a video, told me what it was like to work on a Dole banana farm, when, of course, he is not running or inspiring other fellow workers to take up the sport. Through this form of digital transparency, I now know more about the farm and farm workers who brought me my banana than I do about the farms that I pass by on my way to work, the cashiers I say hello to at my local supermarket or the drivers who sometimes deliver my groceries. In this, digital food activism can bring the distant close and make it visible, but it can, somewhat ironically, make the close seem more distant and *in*visible. As various chapters in this volume point out, transparency is selective and particular in ways that both open up pathways of hope and possibility but also close down others through the relentless application of technology, capitalism and the limits to knowing. As the volume shows, we must be wary and vigilant of the political economies of digital infrastructures and the economies of attention, creativity and geography that produce the online food cyberscapes we can both see *and not see* in the affordances of digital food activism.

Questions still abound of course, some taken up in the volume, others inspired from its contents. First, how does online food activism and their socio-material 'parliament of things' (Latour, 1993) become the durable stuff of the everyday? Put another way, what are the processes by which the ontological experiments of digital food activism work to break us out of our existing habits, choices and engagements with food to facilitate new food ontologies about what is 'right' and 'good' to grow and eat? Habits and tastes, much like our current state of politics more generally, are individually enduring and hard to break away from at a time when the sustainability of the food system is in serious question and new modes of knowing and doing are required for a liveable post-Anthropocene. Second, how does digital food activism re-inscribe inequalities, begin to open up opportunities to flatten them or perhaps do both at the same time? Guthman and Brown (2016) certainly have strong opinions about the ways consumers' digital engagements maintain existing inequalities with farm workers in California. But, their overt dismissal of digital food politics is too simplistic. Rather, digital food activism evinces what I have called the 'relational contingencies' of food (Goodman, 2016). This approach to food acknowledges the need for analysts to not just critically investigate food's progressive *and* conservative cultural political economies, as many chapters do here, but, simply put, the *specific* and *contextualised* impacts of digital food activism matter in *contingent* and *relational* ways worthy of deep and nuanced scrutiny. Finally, how does digital food activism challenge us to not just rethink what food is, as argued here, but also what activism is, can and should be? Does the simple existence of digital food activism, whether or not it is shared, liked and favourited, mean progressive work is going on? Does the mere desire to disturb, change and innovate the food system mean new spaces of digital hope are being formed in their wake (cf. Sexton, 2017)? The manifold, connected online – and offline – types of food politics analysed in this volume require us to not just begin to think about what forms political and material activism can and might take but how this committed activism shapes our collective ability to change our food worlds for the better. In this, *Digital Food Activism* is the perfect starting point for any and all researchers focused on studying not only the digitalisation of food and food activism but this volume is essential for those wishing to analyse food politics at this decisive moment of the Anthropocene.

References

Barnes, C. (2017). Mediating good food and moments of possibility with Jamie Oliver: problematising celebrity chefs as talking labels. *Geoforum*, 84, 169–178.

Dole GmbH (2017). Dole Earth. Available at www.dole.eu/dole-earth/.

Goodman, M. K. (2016). Food geographies I: relational foodscapes and the busy-ness of being more-than-food. *Progress in Human Geography*, 40(2), 257–266.

Goodman, M. K., Johnston, J. and Cairns, K. (2017). Food, media and space: the mediated biopolitics of eating. *Geoforum*, 84, 161–168.

Guthman, J. and Brown, S. (2016). I will never eat another strawberry again: the biopolitics of consumer-citizenship in the fight against methyl iodide in California. *Agriculture and Human Values*, 33(3), 575–585.

Latour, B. (1993). *We Have Never Been Modern*. Brighton: Harvester Wheatsheaf.

Leer, J. and Povlsen, K. (2016). *Food and Media: Practices, Distinctions and Heterotopias*. London: Routledge.

Mol, A. (2008). I eat an Apple. On theorising subjectives. *Subjectivity*, 22, 28–37.

Morgan, K., Marsden, T. and Murdoch, J. (2006). *Worlds of Food: Place, Power, and Provenance in the Food Chain*. Oxford: Oxford University Press.

Sexton, A. (2017). Eating for the post-Anthropocene: alternative proteins, Silicon Valley and the (bio)politics of food security. PhD Thesis, Department of Geography, King's College London.

Acknowledgements

This volume developed out of a one-day workshop entitled 'Digital Food Activism', which took place at St. Hilda's College, Oxford, on 25 November 2015. The workshop was convened by the Oxford Food Governance Group and the Unit of Biocultural Variation and Obesity at the University of Oxford. This was the third workshop we organised to present and discuss our research project, entitled 'Understanding Emerging Forms of Food Consumption: The Role of Information and Communication Technologies (ICTs) in European Food Governance'. This three-year research project was funded by the Oxford Martin Programme on the Future of Food at the University of Oxford, and we are grateful for their generous support.

The aim of the Digital Food Activism workshop was to analyse and assess critically how consumer activists, non-governmental organizations and social entrepreneurs use information and communication technologies to facilitate new or alternative forms of engagement with food. Alongside this, the workshop aimed to examine how media, corporations, academia, and government agencies respond to or mediate these engagements. Building on and extending the rich literature on food activism and critical food studies, the workshop focused on discussions of what happens when food goes digital.

We sincerely thank all presenters for their insightful presentations and their motivation to develop their papers into chapters for this edited volume. Their chapters are joined by invited contributions that bring additional perspectives to bear on the topic. Special thanks go out to our keynote speaker Steve Woolgar (University of Oxford and University of Linköping) for his thought-provoking presentation on food and mundane governance, which has inspired our introductory chapter. We also thank Isabel Beshar and Michelle Pentecost for their help in organising and coordinating the workshop. Finally, we thank all attendees, in this workshop and in our previous workshops, for stimulating our thinking on the issues addressed.

Our sincere thanks also go to our series editor, Mike Goodman (University of Reading), who presented at the Digital Food Activism workshop, for his enthusiasm and support for this book project, as well as for his insightful comments. We also thank the two anonymous readers who reviewed our book proposal for their positive and constructive feedback. At Routledge, we thank our commissioning editor, Faye Leerink, and editorial assistants, Pris Corbett and Ruth Anderson, for their continuous support and for seeing this volume through the commissioning and production process. Finally, we would like to thank Amy K. McLennan, Javier Lezaun, and Farzana Dudhwala for their generous feedback on the chapters we have written for this volume.

We hope this volume will inspire new, multi-disciplinary conversations at the interstices of critical food, social movements, and digital technology research.

1 Introduction

Digital food activism – food transparency one byte/bite at a time?

Tanja Schneider, Karin Eli, Catherine Dolan and Stanley Ulijaszek

Food transparency by digital means

How many food apps do you have on your mobile phone? How often do you share information on food via a tweet or a Facebook post? How often do you see or upload food photos on Instagram? Even if you do not actively post or seek information about food online, it is likely that you have encountered numerous food posts, images, and videos on various online platforms. Food is an increasingly prominent subject of engagement online, from the aesthetics of cooking to the ethics of shopping. In this volume, we contemplate what happens when food, this visceral and enlivening matter, goes digital – and particularly what happens when activism surrounding food moves into the digital domain.

In examining food activism within contemporary 'digital food cultures' (Lupton, forthcoming), this volume focuses on selected websites, mobile phone apps, and social networking platforms that offer digital modes of activist engagement with food. These new platforms provide diverse types of information: where foods or their ingredients are grown and manufactured, which ingredients or nutrients various foods contain, what health or environmental effects these foods are reported to have, and who owns food products or brands. Together, these strands of information reflect the growing interest in, and concern with, questions of food transparency in the context of food-related anxieties among consumers in the Global North (Jackson, 2015).

What is unique about the ways in which food transparency is conceptualised in the digital realm? While food labels carry information on ingredients, nutrients, and safety (e.g. sell-by dates), activists – both offline and online – have pointed out that other consequential information remains hidden or difficult to obtain. Prominent lacunae include the political alliances of food companies and their interconnectedness with other companies, and the environmental impact of various foods (e.g. food miles, or the distance a food travels from its locus of production to its locus of consumption). Into this informational gap step digital food activists, who use the digital realm to redefine and/or expand food transparency, and to disseminate otherwise 'hidden' information to citizen-consumers who may share these concerns.

An example of the process of redefining appears in consumers' use of the mobile app Buycott, which we studied as part of our project on digital food activism (see Eli *et al.*, 2016). Buycott is a US-based barcode scanner app, with a global database that encompasses a range of retail products, including, pre-dominantly, food. The app adopts the term buycott – antonym of boycott – as its name which means to purchase a product or brand deliberately in order to signal support of companies' or countries' policies or practices (Sandovici and Davis, 2010: 329). This concept of political consumption is embodied in Buycott's stated mission to 'vote with your wallet' (Buycott, 2016). Users of the app can generate their own activist campaigns, as well as provide data to inform other users' activist campaigns. While campaigns may focus on any political issue, Buycott includes a number of prominent campaigns concerned with food, includ-ing 'Pro GMO? Or Pro-Right To Know', 'Monsanto Products Boycott – Say Not to GMO', and 'Support Organic Dairy Products' (Buycott, 2017).

In Buycott's case, food transparency is visually depicted through corporate kinship charts (see Eli *et al.*, 2016). When consumers scan a food product with Buycott, the app informs them which parent company owns the product or brand, producing a phylogenic tree of corporate ownership structures. The app also alerts consumers to conflicts between the campaigns they have set up and/or joined and certain food production practices, such as the use of genetically modi-fied organisms (GMO), in which either the company or its parent company engage. When we embarked on our research project, Tanja set out to test a couple of food apps, take notes, and report back on her experiences to colleagues of the Oxford Food Governance Group (the co-authors of this Introduction), which we later shared as an amended version on the group's shared blog.[1] This is an extract from her report (Schneider, 2013) on learning how to buycott using Buycott:

> The first product I scanned was a water bottle that was sitting on my desk right beside me. The water I had bought and was drinking that day uses the name of a Swiss thermal spring in its name and so I had assumed I was buying a local product and supporting a local company located in a Swiss mountain valley. To my surprise, I found out that the water brand was owned by Coca-Cola, a company whose products I don't tend to buy very often. So, unintentionally, I had been spending money on one of their products.
>
> My surprise and my researcher's curiosity led me to check out the water company's website and search for more information on who owns the company. Interestingly, there is no mention that the company is owned by Coca-Cola under the 'company' tab on their website. So I suppose the use of the app was helpful right away in knowing more about the products I consume. However, the information left me wondering what to do now. Being only an occasional drinker of bottled water but a frequent user of empty plastic bottles for tap water refills when I'm on the go, I knew that the clever solution would be to stop buying/using plastic bottles and buy a glass, metal or hard plastic water bottle for refill. That would have the positive side-effect of reducing plastic waste. On the other hand, it would

have the negative side-effect that I wouldn't support jobs in a company located in a Swiss mountain village.

[...]

Reflecting further about my experience with the app, I kept wondering how the app and the information it provided had altered my relationships to the products I buy and consume ...

Here we see how the use of an app like Buycott resignifies a mundane bottle of water, connecting everyday consumption to broader ethical concerns such as local sourcing, corporate profits, and environmental sustainability. As the chapters in this volume suggest, digital food activism is enacted not merely through technologies, foods or 'things', but rather through the relationships between them, as they are mediated and transformed by digital network infrastructures. To study these relationships, we draw on our multi-disciplinary backgrounds in anthropology, sociology and science and technology studies (STS) to address the overarching research question of this volume: *how do diverse actors, including activists, computers, mobile phones, digital network infrastructures and platforms, enact new relationships between food, its producers and consumers, with implications for food activism and food governance?*

With this question in mind, we first review the literature on food activism and digital activism. We then introduce the concept of *digital food activism*, which we have developed to capture diverse forms of digitally mediated practices of food activism, their distinctiveness and their constitutive effects. Next, we situate these practices within the larger multi-disciplinary literature on digital devices, platforms and infrastructures, focusing on the affordances[2] of digital platforms; here, our aim is to explore the kinds of interactions these platforms enable and constrain, and what this means for digital food activism. We then consider digital platforms used for food activism as 'infrastructures that give rise to ontological experiments' (Jensen and Morita, 2015). Building on our own research and that of the contributors to this volume, we show the multiplicity (cf. Mol, 2002) and experimental nature of digital food activism and call attention to how food is ontologically respecified in the entanglements of diverse types of activists and digital platforms. To conclude, we discuss the implications of this ontological respecification for agency, democracy and the economy, and elucidate the similarities and differences between 'traditional' food activism and digital food activism (Counihan and Siniscalchi, 2014).

Digital food activism

This volume stems from our three-year research project,[3] entitled 'Understanding Emerging Forms of Food Consumption: The Role of Information and Communication Technologies (ICTs) in European Food Governance'. When we began our project, we encountered a surprising lack of engagement between extant research in food studies and research in digital sociology, anthropology and geography, as well as communication studies and political science. In response, we developed the concept of digital food activism, to bridge these previously disparate research realms. In developing this concept, we built on existing research concerning food activism and digital activism, as well as on Science and Technology Studies (STS), anthropological and related studies of infrastructure, with a specific focus on digital platforms.

Food activism

In Counihan and Siniscalchi's introductory chapter to their edited volume, *Food Activism*, they define food activism as 'efforts by people to change the food system across the globe by modifying how they produce, distribute, and/ or consume food' (2014: 3). These efforts encompass 'people's discourses and actions to make the food system or parts of it more democratic, sustainable, healthy, ethical, culturally appropriate, and better in quality' (Siniscalchi and Counihan, 2014: 6). They range from spontaneous and institutionalised, to individual and collective acts and are performed by 'political activists, farmers, restaurateurs, producers and consumers' (ibid.: 6–7). *Food Activism* is part of growing scholarship in the field of critical food studies that considers a set of multi-sited practices of ethical consumerism, commodity chains, local food, community-supported agriculture, food-focused social movements, peasant movements for food justice and social movements against biotechnology and agribusiness (ibid.: 4). Grounded in an ethnographic approach to food activism, the volume makes important contributions to our understanding of consumer citizenship and alternative food networks in specific cultural settings (see also Goodman *et al.*, 2014; Grasseni, 2013; Kneafsey *et al.*, 2008; Lien and Nerlich, 2004).

Though *Food Activism* examines multiple forms of food activism, the use of websites, blogs, social media and mobile applications does not feature prominently in the volume. This is surprising, given the usage of websites and blogs by a range of food-focused social movements and alternative food networks (e.g. the Slow Food movement, the Fair Trade movement or Solidarity Purchase Groups) and the frequency with which activists employ social media as a key communication platform. In addition, a review of the wider literature in food studies reveals that examinations of how food production and consumption practices are portrayed and discussed online tend to focus narrowly on consumers' or eaters' food-related social media practices and the roles these play in individual identity formation, rather than on collective, organised activism aiming to change food practices or the food system as a whole. For instance, de Solier (2013) dedicates the final chapter of her book on *Food and the Self* to food blogging and argues that blogging is a meaningful and self-defining practice for foodies in post-industrial times. Blogging, according to de Solier, enables people to share their culinary knowledge (about cooking, growing food or reviewing restaurants) rather than merely to consume or acquire other people's (often food professionals') knowledge about food and eating. Rousseau (2013) illuminates the roles of social media in food procurement, cooking and eating, and the emergence of a digital food community, ranging from food professionals to, in her words, 'food amateurs' who participate in food-centred reviews, displays or conversations through Twitter, Facebook, YouTube, blogs and other formats. Both books provide important insights into the growing uptake of new media and the manifold roles they play in re-mediating food messages or even provoking new food practices (see also Leer and Povlsen, 2016). However, their focus on individualised consumption practices does not elucidate activists' uptake of ICTs to challenge and critique the current food system or propose alternative visions.

Thus far, there has been limited scholarly consideration of food activists' use of digital platforms to challenge, critique and change the conventional global agri-food system. One exception is the research of Alana Mann, who engages with food activism and explores some of the roles that new ICTs play in fostering alternative food networks' goals. In her book, *Global Activism in Food Politics: Power Shift*, Mann (2014) studies the transnational social movement La Vía Campesina, which acts as a global umbrella advocacy organisation for peasant farmers and landless workers. The movement promotes food sovereignty with the goal that those producing, distributing and consuming food should be in charge of the mechanisms and policies of food production and distribution. Focusing on case studies of three member organisations of La Via Campesina located in Chile, Mexico and the Basque country in Spain, Mann analyses how the realities and priorities of local chapters are translated into the global movement's agenda, and in the process evaluates the capacities and limits of transnational advocacy. Mann shows how access to alternative media and digital platforms has become crucial for La Via Campesina and its member organisations in challenging dominant narratives converging around food security – narratives which are driven by corporate actors favouring technological solutions to end global hunger. As an example of the digital contestation of dominant narratives, Mann discusses the online initiative *farmsubsidy.org*, which provides an alternative, virtual sphere where data on farm subsidies in EU states is made publicly available, thus touching on digitally enabled food activism.

Another study that engages with digitally enabled food activism is Cristina Grasseni's (2013) ethnographic research on *Gruppi di Acquisto Solidale* (GAS) (Solidarity Purchase Groups) in Italy, in which farmers and consumers connect directly (without intermediaries such as retailers) and negotiate the terms of producing, selling, buying and preparing food. Grasseni's research reveals that ICTs play a negligible role in these negotiations, in part because many GAS members refuse to place orders online via the organisation's website. These members emphasise the problem and, in their view, the paradox of using global digital network infrastructures to support local food provisioning infrastructures and farmers; they argue that digital platforms are another form of intermediation, albeit a digital one, that blocks direct forms of information exchange and communication. GAS members, then, foreground the value of direct, personal and offline communication to foster the affective relations central to the sort of 'solidarity economy' they seek to foster (ibid.).

In a recent study exploring how alternative food networks (AFNs) draw on online spaces to foster 'reconnection' between producers and consumers, Bos and Owen (2016) emphasise that 'there is scope to better understand the relationship between online space and reconnection, particularly in light of the increasing usage and embeddedness of online and social media activity across society' (2016: 4). While they acknowledge existing studies (Cui, 2014; Fonte, 2013; Holloway, 2002; Reed and Keech, 2017) that have started to explore the online spaces used by AFNs, Bos and Owen (2016) argue that further research needs to be done. Based on their study of eight AFNs in England, they propose that online spaces (e.g. websites, Facebook and Twitter pages) offer 'virtual reconnection' – a supplementary realm for reconnecting producers and consumers. However, they do not identify any of these online spaces as substituting for socio-material reconnections, such as 'the biological and tactile qualities of food (such as smell, touch, taste), as well as the social qualities of reconnection (founded on face-to-face interactions and notions of trust) and the embodied ways of knowing' (ibid.: 12).

In sum, our review of the literature on food activism reveals that, until now, critical food studies scholars have considered food activists' uptake, adoption and use of digital platforms only to a limited extent. We find that the few studies that do attend to the use of digital platforms for food activism focus on platforms that foster individualised forms of self-expression (e.g. food tweets, food blogs), or on how offline AFNs extend and supplement their work by 'going online' and offering additional means of reconnection between producers and consumers. What is still missing is research that explores how digital platforms facilitate the exchange of information and communication on transnational food issues between disparate food consumers and producers (but see Mann, 2014), and how online and offline activism on food issues are interwoven and the economic, political and social implications of these processes. These are the gaps we attempt to address in this book. Pivotal questions that emerge in relation to this are: to what extent do AFNs employ websites, blogs or social media platforms as new information and communication channels, with the potential to recruit new supporters? To what extent do these digital platforms generate new forms of interaction between a range of actors, online and offline, to enable digitally enhanced activism? How do different digital platforms facilitate diverse forms of activist engagement?

Digital activism

This is the first volume to explore the emerging roles of digital platforms in food activism. In so doing, our volume extends debates in critical food studies to the digital sphere. In particular, we engage with literatures on digital activism that trace the divergences of digital and analogue activism (e.g. Bennett and Segerberg, 2012; Gil de Zúñiga *et al.*, 2014; Mossberger *et al.*, 2008). Scholarship on digital activism, however, rarely directs attention to the politics and practices of food production, distribution and consumption, and tends to coalesce around themes of political participation, technological innovation and new forms of citizenship. By grounding digital activism in the mundane negotiations of food provisioning and consumption, our volume offers a new approach in the literature on new media and digital activism.

Digital activism, as Edwards, Howard and Joyce define it, is 'an organized public effort, making collective claim(s) on a target authority(ies), in which civic initiators or supporters use digital media' (Edwards *et al.*, 2013: 10). Digital activists employ digital network infrastructure, such as social networks (most prominently Facebook), blogs, micro-blogs (frequently Twitter), email groups (e.g. Google, Yahoo), video (e.g. YouTube), alongside email and SMS, to achieve political and social change. Searching for a term to describe these practices, Joyce (2010) argues that digital activism – as opposed to, for example, cyber-activism, social media for social change, e-activism or info-activism[4] – best captures the scope of activities and tools. According to Joyce, 'digital activism' is 'exhaustive in that it encompasses *all* social and political campaigning practices that use digital network infrastructure; [and] exclusive in that it excludes practices that are *not* examples of this type of practice' (ibid.: viii).

In recent years, researchers from the fields of political science, media studies, geography and sociology have turned their attention to the uses of digital technologies by activists. Scholars have documented and analysed diverse digital activism campaigns and their democratic potentials and limits. Prominent examples include studies of local and global social movements that rely on digital technologies to instigate, inform, grow and coordinate protests with ambitious goals for social and political change, such as the Arab Spring, Occupy or Black Lives Matter (Bennett and Segerberg, 2012; Howard and Hussain, 2013; Tufekci and Wilson, 2012). Other researchers, however, focus on the uptake of digital technologies for political consumerism, i.e. citizens expressing their opinions about consumer products and seeking to influence business through boycotts and buycotts (Stolle and Micheletti, 2015). Food boycotts and buycotts are potent strategies in political consumerism; increasingly, consumer-citizens turn to digital platforms to search for or share information about food products, as well as about food producers and their commitment to labour rights, ethical sourcing and sustainability (Eli *et al.*, 2016; Humphery and Jordan, in press; Schneider *et al.*, forthcoming).

While studies have provided initial insights on the remit of digital activism, questions about the motivations that underlie digital activism and the predictors of its reach and success remain. To answer these types of research questions, a group of scholars based in the Department of Communication at the University of Washington in Seattle founded the Digital Activism Research Project (DARP)[5] in 2012. DARP's aim is 'to build our collective understanding of the effect of digital technology on political outcomes around the world' (Edwards *et al.*, 2013: 6). To do so, the team created the first longitudinal datasets[6] of digital activism cases worldwide between 1982 and 2012, and developed software to collect additional data. The analysis of the datasets revealed three key findings (ibid.: 4): First, digital activism is civil, non-violent and rarely involves hackers.

Second, Facebook and Twitter dominate global activism but there are many regional digital platforms. Third, the success of digital activism campaigns depends on target type and tool diversity, meaning that campaigns are most successful at galvanising public protest when the government is the target, when the regime is more authoritarian or when the campaign uses multiple digital tools (ibid.: 15). These observations are informative for analysing existing and emerging forms of digital food activism; we return to this in the final section, where we consider the cases of digital food activism presented in this volume and how digital forms and formats mediate food activism.

In our analysis of digital food activism, we examine the overlaps between the food activism literature and the digital activism literature. Though the absence of food-related analyses in extant digital activism literature may seem to imply that digital activism, which focuses on challenging a political-economic order rarely targets food provisioning and governance, our research suggests that food activism plays a central role in the broader political work of non-governmental organisations (NGOs), civil society groups and social entrepreneurs. Like 'analogue' food activists, digital food activists highlight food as highly political, with the production, distribution and consumption of food envisioned as entangled in dominant scientific, agricultural, economic, social, cultural and political dynamics.

Digital food activism

Building on our review of food activism and digital activism, we define *digital food activism as an Internet-based, organised effort to change the food system or parts thereof in which civic initiators or supporters use digital media*. Our definition emphasises the distinction between 'Internet-enhanced' and 'Internet-based' activism (Vegh, 2003). We consider this distinction crucial as digital food activism is shaped by and through the digital media platforms that activists employ. The chapters in this volume highlight the interweaving of digital platforms with the origins, development and implementation of food activist projects. Through the case studies explored in this volume, we see that digital platforms are conceptualised not as *supporting* consumer action, but as *fostering* and *mediating* activism. Thus, examining the affordances of digital platforms is central to the study of digital food activism. In the next section, we review recent STS, anthropological and related studies of digital devices, platforms and infrastructure, and consider what kinds of interactions these enable and what this means for digital food activism.

Digital (food) platforms and ontological experimentation

Media scholar José van Dijck (2013) observes that the rise of social media platforms over the last decade has had profound effects on social interaction. Based on her historically informed study of the 'ecosystem of connective media' and detailed analysis of each of the five leading companies in the social media realm (at the time of her study) – Facebook, Twitter, Flickr, YouTube and Wikipedia – she argues that the development of social media has resulted in an increasingly mediated culture, which she describes as a 'culture of connectivity'. Van Dijck describes this digital connectivity as 'platformed sociality' (ibid.: 5). To consider the conditions and consequences of the new 'platformed sociality', van Dijck analytically disassembles digital platforms:

> By taking apart single platforms into their constitutive components, we may combine the perspectives on platforms as techno-cultural constructs and as organized socioeconomic structures. But disassembling platforms is not enough: we also need to *reassemble the ecosystem* of interoperating platforms in order to recognize which norms and mechanisms undergird the *construction of sociality and creativity.*
>
> (Ibid.: 25, emphasis in original)

This enables van Dijck to study those aspects of a culture of connectivity that are usually backgrounded. Similar to anthropological and STS studies of infrastructure, such an approach foregrounds 'embedded, often invisible technical support structures that help to deliver services to a population or organization' (Niewöhner, 2015: 119). These technical support structures can be conceptualised, then, as 'transient embodiments of social, technical, political, economic and ethical choices that are building up incrementally over time' (ibid.: 119). As such, a historical perspective on social media platforms – like the perspective advanced by van Dijck – enables an analysis that 'emphasizes the partial connections between structure and agency, and inquires into the "how" of connecting and its implications' (ibid.: 119; cf. Star, 1999).

Plantin *et al.* (2016) have recently called for combining the perspectives promulgated in platform studies – as exemplified in van Dijck's aforementioned research – with the perspectives developed in infrastructure studies, within STS and information science. They argue that 'digital technologies have made possible a "platformization" of infrastructure and an "infrastructuralization" of platforms. Articulating the two perspectives highlights the tensions arising when media environments increasingly essential to our daily lives (infrastructures) are dominated by corporate entities (platforms)' (ibid.: 3).

The process of 'infrastructuring' (or infrastructural connecting) has recently been described by Jensen and Morita (2015) as an 'ontological experiment'. In their view, infrastructures are complex, heterogeneous assemblages that 'give rise to ontological experiments because they are sites where multiple agents meet, engage, and produce new worlds. Tracing these transformations is an effort to outline the contours of emergent ontologies' (ibid.: 85). Inspired by Plantin *et al.*'s (2016) study, we aim to extend Jensen and Morita's concept, from infrastructuring to processes of platformisation, and trace how digital platforms give rise to ontological experiments. As Jensen and Morita note, in recent years, the question of how the entities achieve ontological status has gained renewed attention in anthropology and STS (e.g. Gad *et al.*, 2015; Knox and

Walford, 2016; Woolgar and Lezaun, 2013, 2015). For instance, Woolgar and Neyland (2013) emphasise the role of classification schemes that order the onto-logy of people and things. Building on their research of waste management/ recycling, traffic control and management of passenger movement through air-ports, they advocate studying on-going acts of ontological constitution with a focus on how accountability relations are re-distributed, leading to (new) forms of governance (ibid.: 56). The authors draw on the example of the McDonald's hot coffee case – a US product liability lawsuit in which a customer sued McDonald's for serving hot coffee that led to third degree burns after the cus-tomer accidently spilled the coffee. They suggest that the legal respecification of coffee in terms of appropriate temperature resulted in coffee's newly accomp-lished ontological status as safe or unsafe to drink. Ultimately, the authors argue that 'to recognize that coffee can be subject to such ontological segmentation is to recognize that constituting the nature of coffee constitutes its accountable character (that is, that it can take part in the distribution of relations of account-ability)' (ibid.: 55). The potential for ontological respecification poses important questions about how digital platforms used in food activism re-classify and co-constitute food, and the extent to which digital platforms define the actions to be taken within activist networks (see Chapters 2, 4, 8 and 11 in this volume).

In another example of ontological respecification, Hawkins, Potter and Race examine the ontological constitution of bottled water and how drinking bottled water enacts new relations and meanings that may interfere with other kinds of drinking practices (2015: 84). Building on recent publications about 'material politics' (Barry, 2013; Braun and Whatmore, 2012; Marres, 2012), their chapter on anti-bottled water activism[7] is especially valuable for understanding how this form of activism, which has been developing over the past decade, renders bottled water and the practice of drinking bottled water contentious. Activists' challenges of bottled water coalesce around the notion of water being a universal human right. Hawkins *et al.* emphasize, however, that the deliberations around bottled water 'are very situated' (2015: 145). They identify anti-bottled water activism as located in countries where water provision is the provenance of the state, and describe how citizens in countries with no reliable or safe water supply welcome bottled drinking water. In other words, Hawkins *et al.*'s chapter on anti-bottled water activism illustrates how bottled water becomes a 'matter of concern' (Latour, 2005) in some places but not in others and the different 'ways in which the same object can prompt heterogeneous concerns and attract mul-tiple publics according to its different ontological realities as commodity, waste, or something else' (Hawkins *et al.*, 2015: 147). For Hawkins *et al.*, this is not simply a process of moralising bottled water. Studying how bottled water becomes an issue in some places (but not in others) shows how activism – with its strategic intent to disrupt dominant market framings – shifts relations between objects and subjects and contributes to the emergence of a changed reality for bottled water (ibid.: 147–149).

Closer to this volume's focus on the roles of digital technologies in food activism, Knox and Walford (2016) pose the question of whether there is an ontology to the digital. Given our focus on digital food platforms and infrastructures, we are particularly interested in the authors' definition of the term ontology as the 'specific role that material arrangements play in bringing forth social realities' (Knox and Walford, 2016: 2nd paragraph). They argue that studying the ontological dimensions of digital technologies enables researchers 'to pay attention to the unexpected or unforeseen effects of digital technologies and their capacity to disrupt, destabilize, rechannel, unsettle, all the while calling forth different ways of relating' (ibid.: 2nd paragraph).

What new insights do we gain from considering digital food platforms as infrastructures involved in relating and redefining human and non-human actors? Drawing on the analytical power of exploring infrastructures as ontological experiments, in the following section we return to the fieldnote on Buycott shared earlier in this chapter. Through this fieldnote we explore who and what meets, engages and produces which kinds of realities, with and through the digital platform at the core of Buycott.

Buycott as infrastructure/platform for ontological experimentation

The simple act of scanning a water bottle using the Buycott app revealed information hitherto unknown to Tanja. Whereas the water bottle she scanned was Swiss 'local water' for her, the bottle she held in her hands after the scan was geographically local water owned by the multinational company Coca-Cola, headquartered in the United States. As discussed earlier, corporate ownership of products is Buycott's key variable of transparency, as well as the referent of information and comparison. Thus, kinship charts that classify food products in terms of companies and parent companies re-configure corporate ownership as 'the' issue at stake.

With Woolgar and Neyland (2013), we see how this particular format of classifying products orders the ontology of the food product and its producers and consumers (see also Bowker and Star, 1999; Mol, 2013). In the process of uploading information, creating campaigns and/or interacting with Buycott's digital platform via its mobile app, a digital respecification of what food is occurs: *What food is* is mediated by information on corporate ownership, which is enabled by a productive assemblage of food product, barcode, programmers, digital devices, users, digital platform, food manufacturers and others. Similar to labelling a food product as organic, Fair Trade or healthy, food products become digitally 'informed materials' (Barry, 2005; cf. Lezaun and Schneider, 2012). In other words, through the Buycott app, food is reclassified and respecified through relations with complex informational and material environments.

However, it is not only food that is respecified. 'Digital devices', as Ruppert, Law and Savage argue, 'are simultaneously shaped by social worlds, and can in turn become agents that shape those worlds' (2013: 22). As the explorative episode of interaction with Buycott shows, producers and consumers of food can be affected by digital engagement, transitioning to an 'activist' status and habitus. For instance, Tanja's particular food values (to support locally produced and preferably locally owned products), led her to reconsider her choice of bottled water (switching to another brand of bottled water) and, in fact, the consumption of 'bottled' water in general. Additionally, through consumer and producer actions, the digital platform itself and its developers, in turn, also become respecified. For instance, consumers can set up campaigns such as the 'Demand GMO Labelling' or 'Long live Palestine boycott Israel' campaigns, which may or may not reflect the developers' food values. However, as the platform becomes known for hosting such a campaign, its developers will also become associated with the campaign[8] and its entailed values, a process described by Eli *et al.* (2016). Ultimately, accountability relations are redistributed in the process of making food digitally transparent: in the realm of digital food activism, food is governed by (frequently corporate) platforms that respecify foods through reclassification. Consumers who use these platforms and engage in the actions envisioned assume the roles of activists.

Food transparency by digital means, reconsidered

As Barry (2016) observes:

> The ongoing existence of infrastructure depends on the cultivation of consumers and businesses that have an interest, and generate interest through its future existence; transparency appears to offer governments and corporations a way of managing the relation between infrastructures and their publics, which needs to be sustained over time.

Previous research has attended to how 'online platforms' define new 'issue publics', through offering a means of association and action to individuals who would not have connected otherwise (Langlois *et al.*, 2009). This research, however, has also emphasised that digitally enabled activism may lead to volatile and ephemeral action across geographic and political boundaries (Kang, 2012). Digital activism may thus result in 'unlocated citizenship' (van Zoonen *et al.*, 2010), that is, citizenship that is not linked to recognised political institutions. Bennett and Segerberg have described this latter scenario as structured by a logic of 'connective action' (as opposed to the collective action prevalent in social movement organisations) which recognises 'digital media as organising agents' (2012: 752) for 'self-motivated (though not necessarily self-centered) sharing of already internalized or personalized ideas, plans, images, and resources with networks of others' (ibid.: 753). As an organisationally enabled network, Buycott is structured around a logic of connective action rather than collective action. This logic, however, entangles activism and business through the platform's crowd-sourced product information database (Lupton, forthcoming; see also Srnicek, 2017 on platform capitalism), which entwines datafication processes and potential commercial interests. This echoes Langley and Leyshon's conceptualisation of the power of platform-based businesses: power that rests on practices of inter-mediation and processes of capitalisation (2016: 3). Thus, we suggest that an analysis of digital food activism requires careful investigation of what we would describe as 'infrastructures and platforms in the making', which considers how entrepreneurs, digital platforms, consumer-citizens and food are entangled in particular ways, and the economic, political and social implications of this entanglement.

Where does this leave us with respect to digital food activism? We suggest that, to understand digitally produced food transparency, the following questions should be asked: (1) how are activism, expertise and agency defined on each of the platforms?; (2) how do user actions facilitated on each platform enact activist values and identities?; (3) how does the platform infrastructure create new spatialities and materialities of political action? The contributors to this volume have taken up these (and additional) questions, applying them to diverse field sites and objects of inquiry. In the next section, we offer a short overview of each of the chapters that comprise this volume.

The structure of the book

In Chapter 2, Melissa Caldwell introduces the practices of a newly emerging cluster of food activists who seek to change the nature of food and thereby challenge understandings of food. Through their engagement with a range of different tools and technologies, including digital technologies, these self-identified hackers, inventors, entrepreneurs and citizen-scientists set out to disrupt and redefine 'the structures, limits, and possibilities of food' (p. 26).

Based on her ethnographic observations in the United States, Eastern Europe and Russia, Caldwell describes the hackers' modes of inquiry into food and, by extension, into producing and consuming food as playful and experimental. However, the playfulness and experimentation are underpinned by activists' social justice concerns about equity, safety and access to knowledge in the predominant industrial food system. The extension of non-digital food activist concerns to the digital realm is particularly insightful as these practices offer new spaces and materialities for political action, have the potential to invert expert-lay relations and centre on ideas of shared knowledge (rather than patentability). By inviting others to join and identify themselves as makers rather than consumers, these diverse groups develop participatory communities that allow individuals to reclaim their own autonomy, pleasure and sense of personal responsibility. It is to these unexpected effects of digital technologies and their potential to disrupt, challenge and redefine that Caldwell draws attention, while raising broader questions on how these technologies provide alternative ways of relating to food, the environment and people.

Chapter 3, by Amy K. McLennan, Stanley Ulijaszek and Mariano Beguerisse-Díaz, explores the use of Twitter in linking food, eating and diabetes, and considers the ways in which Twitter is used as a platform for food activism. Activism and protest are but two uses among many for Twitter, but along with Facebook, Twitter dominates global digital activism. Although diabetes is fundamentally related to food consumption, diabetes activism on Twitter rarely mentions food. So, rather than dealing with food activism directly, this chapter examines what 'counts' as activism on social media, and highlights different forms that user-generated digital food activism might take.

Through a network analysis of over 2.5 million English-language tweets that contain the term 'diabetes', the authors find that diabetes tweets are posted by users with very different claims to expertise. Users include individuals experiencing diabetes directly; personal trainers advertising their services; companies selling lifestyle products or services; marketing agencies; and hospitals and health agencies attempting to communicate specific health messages. The most common content falls into four thematic categories: health information, news, social interaction and commercial interests. In the social interaction category sits humour, and the authors go on to interrogate the use of jokes in this context.

The biggest diabetes-related jokes on Twitter bring down the mighty health authorities, admit powerlessness in the face of omnipresent big-brand food products, and remind users that the world of food and health is much bigger than their narrow interpretations of it. Governments and researchers are easy targets for humour, as they position themselves as expert authorities, often downplaying or ignoring the effect of corporate lobbies. Diabetes on Twitter generally highlights the uneven balance of power between citizens, governments and organisations that advocate for healthy diets. This humour does not seek to bring about social or political change, but rather demonstrates a resigned and docile acceptance of the status quo. The authors conclude that diabetes tweets form spaces for social interaction and support, and for coordination of collective action and advocacy. This dynamic organisation around a particular idea – such as using humour to highlight consumers' sense of powerlessness – brings to light shared values but does not necessarily connect to action in the non-virtual world.

Sarah Lyon's ethnographic case study, in Chapter 4, shifts our attention from how consumers engage in digitally enabled political consumption to the less recognised, yet increasingly significant role that digital technologies play for producers in agro-food networks. Her case study focuses on how Fair Trade/organic coffee growers in southern Mexico are engaging with digital technologies to gain the visibility and international connections needed to boost their competitiveness in the crowded consumer market of speciality coffee. Their strategies range from adopting online record-keeping to enable field-to-cup traceability, to using the Internet to identify new buyers and service providers, and most notably, to employing social media to share the experience of coffee growing with diverse constituencies, including consumers, buyers and funders.

As Lyon notes, at one level, the adoption of these digital technologies represents producers' efforts to redress the material inequities that have long characterised the coffee commodity chain, allowing them to capture a higher percentage of coffee profits at origin. At another level, it reflects the way that growers are up-ending the 'political ecological imaginary' (Goodman, 2004: 896) of coffee production that Fair Trade deploys to forge an attachment between morally reflexive Global Northern consumers and Global Southern producers. In a process of reterritorialisation, growers are adopting digital means to reassert ownership over their self-image and place-based identities, and to connect to a locally resonant vision of food sovereignty and social justice. In so doing, the adoption of digital strategies by coffee producers opens up new possibilities for southern producers to assert political agency and reframe the broader contours of food politics in the coffee commodity chain.

Whereas Lyon expands the discussion of digital food activism from consumers to producers, in Chapter 5, Katharina Witterhold questions the consumer-activist category itself, arguing for a nuanced typology that accounts for differences between forms of consumer-side activism. Drawing on a study that examines digital consumer citizenship in Germany, Witterhold presents an in-depth comparison of two types of consumer-activists, which she terms 'Green Buycott' and 'Expressive Lifestyle'. Through the case studies of two young women who participated in the study, Witterhold frames the main distinction between the two types as one of consuming information versus producing and sharing it.

While 'Green Buycott' consumers draw on expert information, in both digital and offline spaces, to optimise their own conscious food consumption, 'Expressive Lifestyle' consumers also produce and share information digitally as part of cultivating a political consumer habitus – a mission in which educating others is entangled with one's own development as a conscious consumer. Witterhold, then, suggests that some conscious consumers embody an individualistic orientation to food consumption, whereas others embody a communal orientation; thus, while the two types of conscious consumers may draw on similar sources and make similar food choices, their motivations differ, as does their potential to scale up individual practice into political organisation and action.

Just as Witterhold's chapter challenges the category of 'consumer-activist', in Chapter 6, Ryan Alison Foley questions what forms of engagement constitute digital food activism. Based on her ethnographic work with Luminare, an Italian social cooperative, Foley explores the coop's use of Facebook, showing how social cause and product promotion continuously interweave. Her analysis problematises the boundaries between activism and marketing, and points to the tensions that characterise the retail side of conscious consumption. Luminare's Facebook posts, Foley writes, may seem to be at odds with its mission: whereas Luminare emphasises the values of sustainability, community and social justice, its Facebook posts emphasise products and prices. Moreover, while Luminare operates as a cooperative, its Facebook posts are constructed as 'top-down' marketing messages, with no aim of fostering dialogue with and between citizen-consumers.

Yet, Foley argues, viewed against the backdrop of Luminare's offline activities, the coop's Facebook posts transcend a simple contrast between activism and marketing. Foley suggests that, through using marketing templates in its Facebook posts, Luminare ultimately 'takes advantage of these existing paradigms to maintain their own alternative distribution model, and consequently to support alternative producers' (p. 125). As such, Foley's analysis shows how, in the context of multi-faceted activist organising, the employment of commercial social media for marketing-oriented messaging highlights the sometimes uncomfortable entanglements of activism and consumerism in the realm of food.

The critical repositioning of the use of commercial social media in activist contexts is at the heart of Eva Giraud's analysis of anti-capitalist food activism in Chapter 7. Giraud's analysis is grounded in participatory action research with *Veggies*, a Nottingham-based grassroots initiative which has both catered activist events and produced food activist media, and in a documentary analysis of the *McInformation* campaign, with which Veggies was affiliated, and which grew as a solidarity response to a libel suit filed by McDonald's against two Greenpeace activists.

Tracing the development of activist-generated McInformation and Veggies media in the digital realm over two decades (1996–2016), Giraud highlights the 'frictions' that inhere in the production of activist media. Centrally, these 'frictions' implicate power structures; for example, while anti-capitalist activists aim to generate information free of hierarchies, when this information is disseminated, it implicates an often-elitist expert voice. Such 'frictions', Giraud suggests, have gained an additional dimension with the advent of Web 2.0 and the resulting replacement of activist-produced alternative media with commercial social media platforms, like Twitter, whose commodification of information contrasts with the anti-capitalist ideals around which these activist communities have formed. Yet, Giraud challenges an easy condemnation of the use of commercial social media in food activism, arguing instead for an analysis that highlights 'media ecologies' rather than particular platforms, and thereby explores how activists employ multiple technologies in relation to one another.

Through examining Veggies and McInformation, Giraud shows how an emphasis on 'frictions' can reveal the agentic and creative ways in which activists negotiate the digital realm in order to engage different constituencies, define their identities, and organise effectively within existing structures. In a central example, Giraud highlights how the Veggies website presents the Veggies' Twitter feed alongside links to archived McInformation work, inextricably connecting past and present, commercial and alternative media. Such a direct juxtaposition, Giraud argues, provides a lens onto the ways in which activists 'tinker' with the affordances of digital media to negotiate and arrange enduring activist spaces within 'a shifting media environment' (p. 147).

In Chapter 8, Deborah Lupton and Bethaney Turner consider foods fabricated using 3D printing technologies. Producers of 3D printers for human food production frequently describe their products as one potential solution for challenges such as food sustainability, food waste, ethical consumption, environmental degradation and world hunger issues. These challenges have begun to capture the public imagination through, for instance, increasing media reportage on food security and critical food documentaries, with 3D printing technologies depicted through techno-utopian marketing claims made by scientists, entrepreneurs or hybrid scientist-entrepreneurs.

In their chapter, Lupton and Turner pose a crucial question for the 3D printing industry and its supporters: do these technologies and their resulting food products convince potential eaters? The authors report on an online discussion forum with 30 Australian participants who considered the potential consumption of 3D printed food. Their responses, as the authors describe, vacillated from fascination to disgust. In particular, the discussions centred on the perceived unnaturalness of 3D printed food. Lupton and Turner suggest that those who promote the concept of fabricating food with 3D printers, including activists for sustainability and ethical consumption, need to come to terms with these cultural meanings and dilemmas.

Beyond this advice to activists, scientists and entrepreneurs, Lupton and Turner's chapter reveals how food activists' values have entered research laboratories and start-ups, resulting in a blurring of boundaries between the activist and scientific spheres. We take this as an indication that new spatialities and materialities of political action are taking shape through emerging forms of digital food activism. The most compelling issue that Lupton and Turner introduce is how different actors, including, but not limited to, technology companies, activists and consumers/eaters, redefine what food 'is'.

In Chapter 9, Alana Mann examines the role that digital media and, in particular, Twitter, play in propelling activist action around the issue of food insecurity in Australia. Focusing on the Right to Food Coalition, an Australian food security workforce of policy-makers, academics, community development and health professionals, she suggests that platforms such as Twitter enable activists to bypass journalistic intermediaries and facilitate community organising.

Employing the Twitter Capture and Analysis Tool Set (TCAT), Mann conducts a frame analysis of Twitter coverage of a Right to Food Coalition campaign in Australia leading up to the federal election in 2016. She shows how Twitter elements such as hashtags, handles and mentions can serve as issue-framing devices and mediators within food advocacy networks. Mann's study reveals the importance of Twitter 'for facilitating a cross-flow of information between ideologically aligned advocacy organisations, both domestically and internationally' (p. 169). She discusses how digital media platforms and applications contribute to the constitution of issues and related issue publics – in this case, mostly NGOs. This is an important point that echoes McLennan *et al.*'s observation in Chapter 3 that prominent voices on Twitter are rarely individuals expressing their views, but rather bots or organisations that professionally manage their Twitter accounts.

More generally, this begs the question of whether digital media contribute to a democratisation of public debate or rather to a fragmentation into issue publics assembled around specific, pre-defined issues (by NGOs and other organisations) on a voluntary basis. Although Twitter creates new digital spatialities of political action, the platform and the digital spaces enabled by it are algorithmically structured to foreground those individuals and organisations with many followers, retweets, and strategic and consistent use of hashtags. Mann's chapter, then, suggests that the implicit hierarchy of those with carefully curated Twitter accounts creates a bias towards certain activist values and identities compared to others with fewer resources, such as time, access and digital knowledge.

Tania Lewis explores food politics in a digital era in Chapter 10. Lewis offers an overview of the key challenges that food producers, consumers and activists face. She discusses the growth of lifestyle and consumer-related forms of participatory politics online, the affordances of online platforms for enabling connected forms of personal consumption, and the role of platforms in bringing together food communities. Drawing on her participatory ethnographic research among urban food production activists in Melbourne, Australia, Lewis discusses a Permablitz group whose members redesign suburban gardens based on permaculture principles. She suggests that Permablitz Melbourne 'is not so much a community in the conventional sense as a fluid network of people, connected primarily via a website' (p. 191).

Lewis argues that in the case of Permablitz Melbourne, we can see how the Internet is central for co-producing local food activism. By documenting their efforts online, this group of activists enables others to learn more about Permablitz initiatives and to follow their example by developing their own groups in other places. Lewis also examines the limitations of connectivity and the hegemony of corporate food politics in social media spaces, and considers the limits of digital food activism in an era of Big Data. She highlights that digital food activists dedicated to sustainable food production, distribution and consumption should take into account the potential environmental and social impact of digital infrastructure and energy-reliant communicative systems.

In Chapter 11, Karin Eli, Tanja Schneider, Catherine Dolan and Stanley Ulijaszek return to their research on digital food activism. Focusing on three case studies – a mobile app, a wiki platform, and an online-centric activist organisation – they examine how interactions between activists and platforms generate new knowledges and practices in relation to consumer-based food activism. Specifically, they critically analyse how consumer activists and social entrepreneurs use ICTs to facilitate new or alternative forms of engagement with food, and how platforms, in turn, shape possibilities for action.

Their aim in this chapter is 'to capture diverse forms and potentials of digital food activism, and develop an analytic framework that can be applied to other cases in the field' (p. 203). They propose that digital food activism goes beyond food activism that occurs on digital media. Instead, they argue, digital food activism comprises forms of food activism that are 'enabled and shaped by and through digital media platforms' (p. 203). Digital platforms thus have the potential not only to supplement but also to reimagine AFN activism, creating new messages and activist publics – ideally in interaction with users. However, they acknowledge that many platforms used for digital food activism pre-structure how issues and actions are framed or classified. In addition, they point to the 'free labour' that users provide when they contribute crowd-sourced data on food, and the commercial value such data may have for the owners of digital platforms.

In his Afterword, Javier Lezaun reflects on digital food activism, proprietary versus open source platforms and public participation. Lezaun engages with and accentuates key issues identified by the volume's contributors, and points to the potential of open source digital platforms to create new forms of community that offer space and materials to fashion new publics – publics centred in an activist identity rather than in a consumer identity. Drawing on Kelty's (2008) notion of the 'recursive public', Lezaun suggests that

> Digital food activism thus raises the possibility of a doubly-recursive public, a public that attends both to the conditions under which the food it consumes are produced, and to the systems that generate and disseminate the information that underpins its food choices.

(p. 225)

These infrastructural arrangements and their ensuing ontological experimentation are at the heart of this volume's approach to current debates in food studies and beyond.

What this volume adds

The chapters that comprise this volume call for careful consideration of the ways in which digital media and technologies offer new spatialities and materialities of political, economic and social action concerning food. The experimental nature of some of the activist undertakings the chapters describe is underpinned by an ambition to challenge existing forms of food production, distribution and consumption. The chapters, we suggest, show how digital food activism develops, extends and reimagines food activism, foregrounding 'connective action' rather than 'collective action' (Bennett and Segerberg, 2012). What digital food activism is, however, remains contested. This volume demonstrates that food, and the ways in which it is produced, distributed and consumed, become an issue in some places, for some individuals and for groups, but not in/ for others. The ontological reality of food as political, then, is situated. Producers, consumers and the digital platforms and technologies they draw on enact different food futures, including divergent 'ideal' behaviours and rules to live by, and implicating a diverse 'ontological politics' (Mol, 1999; see also Mol, 2013) of food and the digital.

So, how successful is digital food activism so far? And how successful might it be in the future? Will it pose a major challenge to the industrialised agri-food system and its key players, the large multinational food and agricultural corporations? Will digital food activists foster widespread food transparency? We end on a cautionary yet optimistic note. Despite the potential we see for activists to employ digital media and technologies to reimagine food futures, these efforts are likely to be increasingly monitored, at times by the industry and policy actors targeted. As digital food activism also entails potentially mine-able data about activist groups and consumers, this points to the centrality of infrastructures and platforms in designing effective, and safe, consumer action.

If activists want to avoid 'platformed food sociality', they might need to develop communication infrastructures that act as alternatives to commercial services (see also Giraud, Chapter 7 in this volume). Alternatively, activists relying on proprietary platform and communication infrastructures might achieve visibility through commercial platforms, thereby mainstreaming digital food activism. Another possibility would be for activists to develop spaces, such as the food hacking spaces described in Caldwell's Chapter 2 in this volume, that foster playful and open-ended experimentation with foods. The important thing, we suggest, is that any form and format of collective experimentation entails a broad range of constituencies that come together in 'hybrid forums' (Callon *et al.*, 2011). These public spaces, accessible to diverse groups, including 'experts, politicians, technicians, and laypersons' (ibid.: 18), hold potential for new, multi-vocal provocations and responses that may challenge, and ultimately change, the production, distribution and consumption of food.

On an optimistic note, we suggest that the digital realm could function as a heterotopia (Foucault, [1984] 1986) in which digital technologies would contribute to enacting different food-related realities. Viewing digital food platforms as infrastructures involved in relating and redefining human and non-human actors, these ontological experiments have the potential to reclassify food, shift accountability relations and disrupt prevailing market framings.

Acknowledgements

An earlier version of the argument presented in the first part of this introduction was presented as part of the panel 'Infrastructures, subjects, politics', organised by Jane Summerton and Vasilis Galis at the bi-annual joint conference of the Society for Social Studies of Science (4S) and the European Association of Science and Technology Studies (EASST) in Barcelona, Spain, in September 2016. We would like to thank the panel organisers for this opportunity, and the panel participants and attendees for their helpful feedback. We also thank Farzana Dudhwala and Javier Lezaun for their close reading and valuable comments on this introduction.

Notes

1 https://oxfordfoodgovernancegroup.wordpress.com (accessed 7 July 2017).
2 Affordance is an important analytical concept in several academic fields including STS, media studies and communication studies. Our use of the term draws on Davis and Chouinard's (2017) review of these literatures and the authors' suggestion to attend to the mechanisms and conditions of affordances to consider '*how* artifacts afford, *for whom* and *under what circumstances*'.
3 This chapter is based on research funded by the Oxford Martin Programme on the Future of Food, and we are thankful for the Oxford Martin School's generous support.
4 Joyce explains that '[c]yber-activism refers to the Internet; social media for social change refers to social software applications; e-activism refers to electronic devices' and '[t]he scope of practices encompassed by info-activism is broader than those encompassed by digital activism, so the term is exhaustive but not exclusive' (2010: ix).

5 For more information, see http://digital-activism.org (accessed 10 January 2017).
6 The Global Digital Activism Data Sets (GDADS 1.0 and 2.0), are available at the project website (www.digital-activism.org) and through the Interuniversity Consortium for Political and Social Research (ICPSR) that can be accessed at: www.icpsr.umich.edu/icpsrweb/ (accessed 23 March 2017).
7 According to Hawkins *et al.*, anti-bottled water activism encompasses a diverse set of 'online campaigns, NGO reports, newspaper articles, YouTube videos, public art events, memes and more' (2015: 145).
8 We also discuss Buycott's efforts to be a 'neutral platform', which is visible in the organisation's online Terms and Conditions section that reads:

> Buycott and the Services are about giving Consumers, Campaign Creators and other relevant third parties a voice in relation to socially conscious purchasing. We provide the Services, but the Services are about sharing your voices, not ours. We don't endorse any Campaigns or the Causes to which they are directed. Nor do we have any obligation to screen or vet Campaigns or Causes.
>
> (See www.buycott.com/terms, accessed 25 July 2017)

References

Barry, A. (2005) Pharmaceutical matters: The invention of informed materials. *Theory, Culture & Society*, 22(1): 51–69.

Barry, A. (2013) *Material Politics: Disputes along the Pipeline*. London: Wiley-Blackwell.

Barry, A. (2016) Infrastructure made public. *Limn*, 7. Available at: http://limn.it/infrastructure-made-public/ (accessed 31 April 2017).

Bennett, W. L. and Segerberg, A. (2012) The logic of connective action: Digital media and the personalization of contentious politics. *Information, Communication & Society*, 15(5): 739–768.

Bos, E. and Owen, L. (2016) Virtual reconnection: The online spaces of alternative food networks in England. *Journal of Rural Studies*, 45: 1–14.

Bowker, G. C. and Star, S. L. (1999) *Sorting Things Out: Classification and Its Consequences*. Cambridge, MA: MIT Press.

Braun, B. and Whatmore, S. (eds) (2012) *Political Matter: Technoscience, Democracy and Public Life*. Minneapolis, MN: University of Minnesota Press.

Buycott (2016) Vote with your wallet. Available at: www.buycott.com (accessed 15 July 2017).

Buycott (2017) Campaigns. Available at: www.buycott.com/campaign/browse (accessed 15 July 2017).

Callon, M., Lascoumes, P. and Barthe, Y. (2011) *Acting in an Uncertain World: An Essay on Technical Democracy*. Cambridge, MA: MIT Press.

Counihan, C. and Siniscalchi, V. (eds) (2014) *Food Activism: Agency, Democracy and Economy*. London: Bloomsbury.

Cui, Y. (2014) Examining farmers markets' usage of social media: An investigation of a farmers market Facebook page. *Journal of Agriculture, Food Systems, and Community Development*, 5(1): 87–103.

Davis, J. L. and Chouinard, J. B. (2017) Theorizing affordances: From request to refuse. *Bulletin of Science, Technology & Society*, First published 16 June 2017.

De Solier, I. (2013) *Food and the Self: Consumption, Production and Material Culture*. London: Bloomsbury.

Edwards, F., Howard, P. N. and Joyce, M. (2013) Digital activism and non-violent conflict. Seattle: University of Washington. Available at: http://digital-activism. org/2013/11/report-on-digital-activism-and-non-violent-conflict/sthash.PodOSCHx. dpuf (accessed 1 May 2017).

Eli, K., Dolan, C., Schneider, T. and Ulijaszek, S. (2016) Mobile activism, material imaginings, and the ethics of the edible: Framing political engagement through the Buycott app. *Geoforum*, 74: 63–73.

Fonte, M. (2013) Food consumption as social practice: Solidarity purchasing groups in Rome, Italy. *Journal of Rural Studies*, 32: 230–239.

Foucault, M. ([1984] 1986) Of other spaces, trans. J. Miskowiec. *Diacritics*, 16(1): 22–27.

Gad, C., Jensen, C. B. and Winthereik, B. R. (2015) Practical ontology: Worlds in STS and anthropology. *Nature Cultures*, 3: 67–86.

Gil de Zúñiga, H, Copeland, L. and Bimber, B. (2014) Political consumerism: Civic engagement and the social media connection. *New Media & Society*, 16(3): 488–506.

Goodman, D., DuPuis, M. E. and Goodman, M. K. (eds) (2014) *Alternative Food Networks: Knowledge, Practice, and Politics*. London: Routledge.

Goodman, M. K. (2004) Reading fair trade: Political ecological imaginary and the moral economy of fair trade foods. *Political Geography*, 23(7): 891–915.

Grasseni, C. (2013) *Beyond Alternative Food Networks: Italy's Solidarity Purchase Groups*. London: Bloomsbury.

Hawkins, G., Potter, E. and Race, K. (2015) *Plastic Water: The Social and Material Life of Bottled Water*. Cambridge, MA: MIT Press.

Holloway, L. (2002) Virtual vegetables and adopted sheep: Ethical relation, authenticity and internet-mediated food production technologies. *Area*, 34: 70–81.

Howard, P. N. and Hussain, M. M. (2013) *Democracy's Fourth Wave? Digital Media and the Arab Spring*. Oxford: Oxford University Press.

Humphery, K. and Jordan, T. (in press) Mobile moralities: Ethical consumption in the digital realm. *Journal of Consumer Culture*.

Jackson, P. (2015) *Anxious Appetites: Food and Consumer Culture*. London: Bloomsbury.

Jensen, C. B. and Morita, A. (2015) Infrastructures as ontological experiments. *Engaging Science, Technology and Society*, 1: 81–87.

Joyce, M. (ed.) (2010) *Digital Activism Decoded: The New Mechanics of Change*. New York: International Debate Education Assocation.

Kang, J. (2012) A volatile public: The 2009 Whole Foods boycott on Facebook. *Journal of Broadcasting & Electronic Media*, 56(4): 562–577.

Kelty, C. (2008) *Two Bits: The Cultural Significance of Free Software*. Durham, NC: Duke University Press.

Kneafsey, M., Cox, R., Holloway, L., Dowler, E., Venn, L. and Tuomainen, H. (2008) *Reconnecting Consumers, Producers and Food: Exploring Alternatives*. London: Bloomsbury.

Knox, H. and Walford, A. (2016) Is there an ontology to the digital? *Cultural Anthropology*. Available at: https://culanth.org/fieldsights/818-is-there-an-ontology-to-the-digital (accessed 8 May 2017).

Langley, P. and Leyshon, A. (2016) Platform capitalism: The intermediation and capitalisation of digital economic circulation. *Finance & Society*: Early View, 1–21.

Langlois, G., Elmer, G., McKelvey, F. and Devereaux, Z. (2009) Networked publics: The double articulation of code and politics on Facebook. *Canadian Journal of Communication*, 34(3): 415–434.

Latour, B. (2005) From Realpolitik to Dingpolitik or how to make things public. In B. Latour and P. Weibel (eds) *Making Things Public: Atmospheres of Democracy.* Cambridge, MA: MIT Press, pp. 14–41.

Leer, J. and Povlsen, K. K. (eds) (2016) *Food and Media: Practices, Distinctions and Heterotopias.* London: Routledge.

Lezaun, J. and Schneider, T. (2012) Endless qualifications, restless consumption: The governance of novel foods in Europe. *Science as Culture*, 21(3): 365–391.

Lien, M. and Nerlich, B. (eds) (2004) *The Politics of Food.* London: Berg.

Lupton, D. (forthcoming) Cooking, eating, uploading: digital food cultures. In K. LeBesco and P. Naccarato (eds) *The Handbook of Food and Popular Culture.* London: Bloomsbury.

Mann, A. (2014) *Global Activism in Food Politics: Power Shift.* New York: Palgrave Macmillan.

Marres, N. (2012) *Material Participation: Technology, the Environment and Everyday Publics.* Basingstoke: Palgrave Macmillan.

Mol, A. (1999) Ontological politics. A word and some questions. In J. Law and J. Hassard (eds) *Actor Network Theory and After.* Oxford: Blackwell Publishers, pp. 74–89.

Mol, A. (2002) *The Body Multiple: Ontology in Medical Practice.* Durham, NC: Duke University Press.

Mol, A. (2013) Mind your plate! The ontonorms of Dutch dieting. *Social Studies of Science*, 43(3): 379–396.

Mossberger, K., Tolbert, C. J. and McNeal, R. S. (2008) *Digital Citizenship: The Internet, Society and Participation.* Cambridge, MA: MIT Press.

Niewöhner, J. (2015) Anthropology of infrastructures of society. In J. D. Wright (ed.) *International Encyclopedia of the Social & Behavioral Sciences.* 2nd edn. Oxford: Elsevier, pp. 119–125.

Plantin, J-C., Lagoze, C., Edwards, P. N. and Sandvig, C. (2016) Infrastructure studies meet platform studies in the age of Google and Facebook. *New Media & Society.* Available at: https://doi:org/10.1177/1461444816661553 (accessed 13 October 2017).

Reed, M. and Keech, D. (2017) Gardening cyberspace: Social media and hybrid spaces in the creation of food citizenship in the Bristol city-region, UK. *Landscape Research.* Available at: http://dx.doi.org/10.1080/01426397.2017.1336517 (accessed 13 October 2017).

Rousseau, S. (2013) *Food and Social Media: You Are What You Tweet.* Plymouth, UK: AltaMira Press.

Ruppert, E., Law, J. and Savage, M. (2013) Reassembling social science methods: The challenge of digital devices. *Theory, Culture & Society*, 30(4): 22–46.

Sandovici, M. E. and Davis, T. (2010) Activism gone shopping: An empirical exploration of individual-level determinants of political consumerism and donating. *Comparative Sociology*, 9(3): 328–356.

Schneider, T. (2013) Learning how to buycott? Political consumerism and new media. *Oxford Food Governance Group blog (OFG).* Available at: https://oxfordfoodgovernancegroup.wordpress.com/2013/08/20/learning-how-to-buycott-political-consumerism-and-new-media/ (accessed 30 May 2017).

Schneider, T., Eli, K., McLennan, A. K., Dolan, C., Lezaun, J. and Ulijaszek, S. (forthcoming) Governance by campaign: The co-constitution of food issues, publics and expertise through new information and communication technologies. *Information, Communication and Society.*

Siniscalchi, V. and Counihan, C. (2014) Ethnography of food activism. In C. Counihan and V. Siniscalchi (eds) *Food Activism: Agency, Democracy and Economy.* London: Bloomsbury, pp. 3–12.

Srnicek, N. (2017) *Platform Capitalism*. Cambridge: Polity Press.

Star, S. L. (1999) The ethnography of infrastructure. *American Behavioral Scientist*, 43(3): 377–391.

Stolle, D. and Micheletti, M. (2015) *Political Consumerism: Global Responsibility in Action*. Cambridge: Cambridge University Press.

Tufekci, Z. and Wilson, C. (2012) Social media and the decision to participate in political protest: Observations from Tahrir Square. *Journal of Communication*, 62(2): 363–379.

van Dijck, J. (2013) *The Culture of Connectivity: A Critical History of Social Media*. Oxford: Oxford University Press.

van Zoonen, L., Visa, F. and Mihelja, S. (2010) Performing citizenship on YouTube: Activism, satire and online debate around the anti-Islam video *Fitna*. *Critical Discourse Studies*, 7(4): 249–262.

Vegh, S. (2003) Classifying forms of online activism. In M. McCaughey and A. M. D. Ayers (eds) *Cyberactivism: Online Aactivism in Theory and Practice*. London: Routledge, pp. 71–95.

Woolgar, S. and Lezaun, J. (2013) The wrong bin bag: A turn to ontology in science and technology studies? *Social Studies of Science*, 43(3): 321–340.

Woolgar, S. and Lezaun, J. (2015) Missing the (question) mark: what is the turn to ontology? *Social Studies of Science*, 45(3): 462–467.

Woolgar, S. and Neyland, D. (2013) *Mundane Governance: Ontology and Accountability*. Oxford: Oxford University Press.

2 Hacking the food system

Re-making technologies of food justice

Melissa L. Caldwell

What happens to the performative aspects of food preparation and consumption when they move outside the garden or kitchen to a laboratory, the stage, or a virtual world? What happens when food is brought to consumers not by farm-workers, chefs, and delivery persons, but by computer engineers, robots, 3D printers, and drones? What happens to sensory experiences, especially pleasure, when they shift registers from being purely bodily experience to intellectual experiences mediated by computer programs?

As digital technologies (e.g. hardware, software, and mediated online social fora) have created new possibilities for addressing food justice concerns about access, equity, safety, and transparency, they have also revealed new uncertainties about privacy, surveillance, accountability, and bodily integrity. For instance, nutrition and wellness programs that encourage participants to log their dietary intake and exercise via online apps that then allow them to share their data with friends, fellow participants, or health care providers through public or semi-private social media raise critical questions about the nature of communal oversight or whether sharing invites encouragement or shaming. Similarly, digital tracking systems that chart farm-to-table trajectories of food may simultaneously reveal and obscure not only the individuals whose labours have made those trajectories possible (e.g. farmers and harvesters versus technologists), but also the criteria by which safety and healthfulness are evaluated (e.g. transparency in process versus timeliness of delivery). Perhaps most notably, widespread adoption of digital technology overlooks the fact that the costs and mechanics of that technology may in fact be exclusionary. Not everyone can afford digital technology or have access to reliable Internet or cellular phone coverage, and inclusive universal design features have been outpaced by the speed of technological innovations.

Proponents and activists for reforming the food system have approached these competing concerns in different ways that have placed into opposition competing goals about the greater good of the community versus the needs and concerns of any one individual. In some cases, these oppositions are apparent in the attempts to promote knowledge as collective property, as evident in such diverse projects as Yelp reviews of restaurants (Pennell, 2016), recipe sharing via social media (Rousseau, 2012), or community-oriented online weight-loss initiatives

(Contois, 2017). In other cases, these concerns are manifest in projects that encourage consumers to privilege ethical dimensions of production ahead of qualities such as taste, desire, or even pleasure (Lyon, 2014). Connecting these approaches, however, is an interest in reforming and replacing existing food systems through a combination of societal and individual changes that are made possible by digital technologies.

Within this context, another group has emerged. Loosely connected through a shared ideal of the disruptive potential of technology, these individuals are using digital and other technologies to change both the nature and the experience of food. Identifying themselves through such varied names as hackers, inventors, entrepreneurs, makers, DIY-ers, and citizen-scientists, these individuals are upsetting and redefining the structures, limits, and possibilities of food. Specifically, these individuals draw on the simultaneously democratic and anarchist principles of hacking, a movement with its roots in computer programming, to respond to what they perceive as the more hegemonic, exclusionary, and even elitist nature of many food justice movements and, instead, to create more participatory modes of engagement (e.g. Crossan *et al.*, 2016; Kera *et al.*, 2015). In so doing, they are positioning themselves as a different community of food activists, who are focused on disruption as a means to carve out new spaces and opportunities for social justice concerns.

This chapter analyses the ways in which justice-related themes such as democratic participation, freedom, equity, diversity, transparency, and choice play out through different forms of digital technologies, with particular attention to the ways in which individuals engaged in disruption are changing the discourses and practices of food justice. Of particular concern is the nature of 'disruption' as a practical philosophy and how this practical philosophy changes the ways in which food justice problems are identified, imagined, and solved. This chapter first discusses the role of digital technology in food-oriented social progress initiatives, before moving to a consideration of the ways in which digitally enhanced food projects elicit both advantageous and disadvantageous qualities. It then analyses several examples of food disruption and shows the ways in which disruptors' efforts are oriented towards changing the relationships presumed to inhere between problems and solutions. This material draws on on-going ethnographic research on disruptive food practices, including design-influenced ethnographic research among food hacking communities.[1]

Technology and social progress

Food has long been a special target of technological developments, particularly in relation to efforts to improve efficiency and quality, beginning with growing and harvesting and continuing through the food chain to production, distribution, consumption, and more recently, post-consumption evaluation. Canning, refrigeration, conveyor belts, self-cleaning processes, mechanized feeding systems for farm animals, and automated milking systems for dairy cows have all been part of a larger set of initiatives geared at increasing productivity, reducing spillage and waste, and raising standards of health and safety for workers, consumers, animals, and crops.

More recently, digital technologies have become ever more common and expected within food systems at every level. From the most personal and intimate activities of cooking and eating to global industrial agriculture and food manufacturing systems, digital technologies have enabled new methods and techniques for growing, preparing, distributing, and disposing of food. In particular, in a contemporary period marked by increased concern with information, digital technologies offer new ways for producers, consumers, and observers alike to engage with and understand the food system (Gottlieb, 2015; Pennell, 2016; Rousseau, 2012), particularly in terms of assuring safety and transparency.

For instance, in response to the many instances of food scares that occurred across the United States and elsewhere in the 1990s and early 2000s, a common response among food manufacturers was to introduce tracking systems to monitor and regulate food at every moment in the food system. Animals and crops are often tagged with RFID chips that are then scanned and entered into database systems to track them as they move through the food system. Intelligent design and Internet of Things (IOT) processes have expanded the functionality of RFID chips to include detailed information about temperature controls and location, among many other factors.

Manufacturers, regulatory officials, and consumers can now be assured of the origins of their foods as well as their timeliness and security throughout the various stages of the manufacturing and distribution process. When problems arise, they can easily be tracked back to a particular moment and place in the food chain: in transit, in the factory, in the feedlot, or on the farm. Problems can also be traced back to a particular individual – a grower, processor, packager, or driver – thus ensuring that blame and retribution can be personalized.

At the same time, even as these measures were introduced primarily to address problems in the global capitalist food system, they have also become incorporated into food movements that are focused on issues more commonly associated with the antithesis of the global capitalist, commercial food system: local, artisanal, and heritage foods. Like many protected heritage food products, Spanish *jamón* is tagged with chips and barcodes to ensure and prove authenticity, place of origin, and even pig of origin. In a context such as the European Union, where heritage foods and concerns with *terroir* are entangled in suprastate issues of economics and politics, chips and codes designate both the legal and national status of particular objects. This, in turn, protects both consumers, who want to be reassured of consuming 'authentically local' food products, and producers, who need those official guarantees for their own financial benefit.

Beyond matters of authenticity, safety, and comfort, digital technologies can also provide consumers with information about the ethical commitments in which their foods are embedded while also imparting a sense of entertainment, fun, and pleasure. Manufacturers can be playful with the barcodes attached to their products in ways that create more aesthetic forms of foods and food experiences. Manufacturers can also use barcodes to encourage consumers to engage more directly and fully with their products. One corporation that has explored this is Frito-Lays, which has included active barcodes that consumers can scan and then follow via cyberlinks to more information: recipes, nutritional information, and even entertainment. One line of Lays chips invites consumers to follow the barcode to an informational video that is a documentary of the farm-to-shelf-to-table life cycle of the chips. The farmer who grows the potatoes, the person who picks the potato, the driver who transports the potato to the processing facility, the line supervisor, and other individuals involved at different stages of the process are all featured in order to give a sense of the entire community at work behind the chips contained in the bag in the consumer's hand. Another Lays product, Doritos, also features barcodes that allow consumers to link to exclusive music videos and other fan communities. Lastly, bar odes can even communicate visually the ethical stances of the producers – for instance, the animal or human safe labour and transit practices that the producers follow, such as with a World Wildlife Federation (WWF) barcode that promises no wild animal commerce with a picture of a bleeding zebra.

In different contexts, digital technologies are enhancing animal welfare, such as the dairy industry. Dairy cows are tagged with electronic chips or collars that can be scanned by sensors and then allow them access to individualized milking stalls. Once inside the milking stall, sensors measure everything from their milk output to the amount of grain they consume. It can also allow a farmer to pre-set the amount of grain the cow eats, thus enabling a farmer to fatten up or slim down cows as necessary. As such, these new technologies allow for better oversight of cows' health and welfare. But at the same time, they also allow cows to exercise a form of agency. It is up to the cows to approach the stall, thus determining when they get fed. In some cases, cows have figured out how to work the system to get more food or to re-enter the stalls. At the same time, digital technologies have transformed the workspaces and work lives of dairy farmers, so that they are as likely to be found in an office using a computer as they are in a milk barn.[2] Digital tracking technologies have thus become ubiquitous as forms of information, guarantees of security and safety, and techniques for transparency and authenticity.

Additionally, as both social media and digital photography and videography have moved beyond fetishization of culinary aesthetics (i.e. 'food porn'), they have become platforms through which producers and consumers share information about sourcing, ingredients, production processes, and experiences (McBride, 2010). As Michael Pennell (2016) has argued in his case study of the role of Twitter among American chefs and food vendors, social media has increased both transparency and diversity by allowing producers to make visible, often in real-time, the production and distribution processes that are otherwise concealed from consumers. Kitchen appliances are now enabled with Wi-Fi and Bluetooth and can be programmed to send out messages via social media platforms about the state of their contents, and care keepers can monitor elderly or disabled persons through virtual communication interfaces embedded in electric kettles or other small kitchen devices. At the same time, digital technologies have become critical parts of business operations, especially in terms of branding, marketing, and advertising (Gottlieb, 2015; Pennell, 2016). On the other hand, digital technologies are not simply reaching new audiences, but also enabling the differentiation and personalization of those audiences. For instance, by tracking the move of Weight Watchers into cyberspace, Emily Contois has observed that digital technologies have allowed the weight-loss company to attract new constituencies, especially among men, and individualize their programmes through separate online communities for men and women (Contois, 2017).

The effects of greater access to food knowledge also extends to food access more generally, as increasingly, digital technologies enable consumers to select, pay for, and even review groceries and meals through online platforms, including most notably through their smartphones. In the United States, both members in CSA (community-sponsored agriculture) programmes and customers of meal delivery companies like Blue Apron can order their boxes online. Meanwhile, the human face of a delivery person is increasingly replaced by drones that drop off meals and groceries (Spiegel, 2014). Food and beverage stores like Starbucks have installed separate pick-up counters for customers who have pre-ordered and paid for their beverages online. By April 2017, Starbucks was reporting that one-third of its sales the previous quarter had been made online or through its digital app (Wattles, 2017).

Discussions about the new social phenomena created by digital technologies reveal intriguing debates about their capacity for promoting qualities such as healthfulness, safety, and affordability, as well as more explicitly political ideals about sustainability, accessibility, transparency, and equity (Choi and Graham, 2014; Lyon, 2014; McClements *et al.*, 2011). Amit Zoran and Marcelo Coelho have observed that 'an information-driven food culture can be the source of conscious and healthier choices that take place at an individual level and in concert with environmental and global concerns' (2011: 428). More concretely, McClements *et al.* have argued that 'advances in technology and social organization have reduced the costs, increased the availability, and broadened the diversity of all sorts of products, including foods' (2011: 76).[3]

Yet critics have suggested that valorizations of new technologies are predicated on the presumption that these new forms will solve problems (Choi and Graham, 2014; Davies, 2014) without recognizing that social media, virtual communities, and other forms of digitally mediated cultural practices are unsettling other dimensions of daily life by introducing new forms of distance, alienation, or even invisibility. Cyberspace threatens to alienate people from the presumably more intimate qualities of face-to-face social encounters, such as when Weight Watchers displaces its more traditional forms of in-person weigh-ins and counselling sessions to virtual check-ins and videotaped motivational materials (Contois, 2017). Contois reports that when Weight Watchers' system of calculating food points moved to barcode scanners on mobile phones, dieters – men, in particular – began treating them like games rather than as metrics of biomedical health (ibid.: 39–40). New technologies also make possible the denaturing of foods and food experiences, such as with molecular gastronomy

(Roosth, 2013; This, 2006), food fabrication through 3D printing (Sun *et al.*, 2015), and food substitutes such as Soylent (Dolejšová, 2016a; Kao, 2016).

Attention to the duality implicit in digital technologies illuminates their complicated role as arbiters and conduits of moral and ethical issues. Often obscured in promotions of digital technologies and the new forms of knowledge and self-improvement they make possible are their implications for other critical issues such as personal privacy, autonomy, accountability, freedom, and bodily integrity. In the case of the Fitbits and social media sharing of personal activities promoted by health care providers and ordinary people alike, the value of the knowledge, encouragement, and accountability made possible through those digital media is juxtaposed against the issue of turning over intimate knowledge of my body and behaviour to another person and anonymous monitoring system. Where is the boundary between sharing and surveillance? What happens to bodily integrity and autonomy when it is ceded and displaced to external entities?

Tracking devices do not just provide information and assurances about the quality and nature of food products; they also directly link practices and products with individuals who can be identified, and by extension, shamed, blamed, or otherwise made responsible. The most personal aspects of people's daily lives are not just revealed and documented but turned into forms of data that can be used for other purposes, whether those are commercial or not-for-profit, even civically oriented. By following a barcode to a music video or recipe chat room, or by posting reviews on social media, consumers identify themselves to companies and marketers who can then target them with future advertisements and offers. As a result, consumers unknowingly invite marketers and manufacturers into the most intimate spaces of their personal lives and bodies.

By extension, despite the intimacies and immediacies afforded by new digital technologies, distancing also occurs, as producers and consumers are detached from face-to-face contacts, their own bodies, and even their own labour.[4] In the case of digitally enhanced milking parlours, cows are not only alienated from the farmers who care for them, but they are also alienated from their own labour. A *Der Spiegel* article warned that with changes in European Union policy regarding milk quotas, alongside the disappearance of small dairy farms, farmers would need to increase production just to survive. As a result, the article suggested, new technologies will make it easier for farmers to increase milk production – thus turning their cows into machines that exist far away from the meadows (Stock and Würger, 2014; see also Overstreet, n.d.).[5]

Food activists and social reformers have approached these competing concerns in different ways. In some cases, responses have emphasized the greater good of the whole over the needs and concerns of any one individual. This is the case in health-related matters, where activists and social observers promote personal health as a form of collective property that contributes to the well-being of the community and environment. This ideal of the greater social good also appears in perspectives that promote personal choice as something that is logically monitored and determined by collective consensus, such as when consumers are advised of the ethical and environmental implications of the foods they consume (e.g. Biltekoff, 2013; Carney, 2015; Guthman, 2011).

In this context, another group of food activists has emerged. But their approach is primarily focused on the challenges and limits of existing food technologies, especially for critical social justice concerns of autonomy, independence, and equity. Most notably, these food activists are using new technologies to change the nature of food itself – they are deliberately focused on disrupting the food system and transgressing and remaking expectations about food, technology, and social change. In the following sections I describe a series of food-themed events focused on experimentation, transgression, and disruption. I begin with an encounter in Moscow that occurred at the very beginning of an aesthetic culinary movement that has subsequently come to be included in the larger movement of food-related transgression and disruption, before moving to more recent events that have been explicitly described as forms of disruption.

Anticipatory transgressions: from what is to what if?

My summer 2007 fieldwork in Russia coincided with a culinary cultural trend sweeping the country. As was occurring in other global settings where chefs and food enthusiasts were inspired and intrigued by developments in food media, Russia's media were witnessing an explosion of food-themed programming. Both state-owned and private cable television channels featured numerous cooking shows for an increasingly differentiated set of audiences: young, well-to-do, urban hipsters; working professionals with young families; middle-aged men; and frugal retirees who had grown up under the Soviet system. News programmes, daytime celebrity 'talk' shows, survival-themed reality shows, and outdoors-themed television shows began including segments on food acquisition, preparation, presentation, and consumption. Food magazines for sale in the local news kiosks increasingly included CDs of recipes and links to supplemental online resources.

At the same time, home-based Wi-Fi service and smartphones had become mainstream and accessible to many ordinary Muscovites, making digital connectivity an ever-present reality across the city. For those Muscovites who were interested in food, online social media fora offered like-minded individuals opportunities to gather together to share meals, either together or by passing out food to homeless persons, as well as to disseminate information and enable access to an underground restaurant scene.

In the midst of those dynamics that summer, two Moscow radio personalities hosted a series of themed dinners in a hip, upscale, sports-bar-themed brasserie-style restaurant located in the centre of the city.[6] Each event in this particular themed dinner series featured a local chef from another restaurant who had been invited to prepare the dishes and then discuss them with the audience. The explicitly performative mode of the dinners was apparent in the forms of presentation. Diners were seated at tables arranged in front of a small stage where a DJ and musicians provided both taped and live music. Brightly coloured lights flashed overhead, alternately illuminating the presenters and audience, while a glitzy PowerPoint presentation played on a screen mounted above the stage. It was a multi-media spectacle that seemed more like a nightclub show than a dinner. Images of food and information about the foods and food experiences that were projected via the PowerPoint presentation accompanied the hosts' commentaries, while individual guests used their mobile telephones and digital cameras to take pictures of the foods and event.

The two dinners that I attended were structured as part infomercial and part tasting session, as the presenters switched between descriptions of ingredients and 'tasting notes' intended to guide diners on how best to sample and appreciate the dishes, on the one hand, and demonstrations of appliances and cooking techniques, on the other. Laptops jostled for space among kitchen equipment on the presentation table. Apparent in the commentaries of the presenters was the presumption that diners were unfamiliar with such kitchen tools as hand blenders and cast-iron cookware and imported produce, such as kiwis and avocados.

Most notably, the events represented what has since come to be recognized as transgressive or disruptive food experiences, as the food, activities, and hosts' commentaries challenged diners' expectations about acceptability and normalcy. With a deliberate focus on detaching foods from their usual forms and roles, the practical and sensory aspects of eating and drinking were also disrupted, and expected rules about who, what, and how were deliberately reworked or ignored.

Even without knowing precisely the hosts' intentions, the structure and effects of these dinners were very much in line with what has since come to be known more generally as bioart, reflecting what Allison Carruth has called 'experiments in *food culturing*' (2013: 89). Situated at the intersection of food, science, art, and hands-on experiences, bioart presents deliberate disruptions and restructurings of food experiences, often as intentional responses to conventional and dominant food systems (ibid.).

These disruptions were best captured in the dinner entitled 'Immoral Foods', when the hosts directly challenged participants to rethink their moral ideologies about gender, race, and nationality. The 'Immoral Foods' dinner began with a commentary on 'moral beverages'. By way of elaboration on the idea that in many societies beverage consumption is gendered, the hosts discussed absinthe while servers dispensed drinks in cocktail glasses to diners in the packed room. Women received green, foamy drinks in daiquiri glasses, while men received layered green cocktails in martini glasses. Before we were allowed to sample our beverages, the hosts sternly informed us that we were to drink only what was in front of us. Nevertheless, many diners ignored the 'rules' and shared each other's drinks, and the conversations among the couples seated at my table revealed that participants were confused about the gendered moralities they were supposed to associate with the two drinks.

The next course was a 'soup course', which focused on American food cultures and a distinctively American brand of canned soups. By way of introduction, the hosts stated that one of the most significant differences between Americans and Russians was that Americans' knowledge of making soup is limited to being able to open cans, while Russians were such expert soup makers that they do not know what canned soup is or even how to use a can opener. To illustrate this point, the hosts invited onto the stage several audience members whom they 'taught' how to open cans of soup, followed by directions on preparing variations on tomato soup. One audience member prepared a cold gazpacho-style soup by dumping a can of uncooked tomato soup into a bowl and adding parsley, heavy cream, and ice cubes. A second person supplemented a can of tomato soup with chopped leeks and water. The third person prepared a tomato soup by thinning it with water and adding chopped red and yellow bell peppers and a dollop of vanilla ice cream. After the guests had prepared the three soups, servers distributed small samples to the diners, who seemed largely unimpressed.

The next segment of the dinner focused on morality and poverty. As the hosts explained simplistically to the audience, food could be used to satisfy poor people. Specifically using the phrase '*gumanitarnaia pomoshch*' (humanitarian assistance), one host referenced Africa as a segue to a lengthy account of Russian forms of humanitarian assistance to African countries.[7] Following this, the hosts presented an equally long commentary on the plainness and meagreness of a narrowly defined 'African diet' that was composed primarily of banana leaves, chillies, and peanuts – a claim that was expressed in the unsavoury and visually unappealing nut logs wrapped in pieces of waxed paper meant to emulate banana leaves that were served to every guest.

The next topic focused on Colombia and alternative forms of pleasure and energy. In front of every two participants, servers placed several small straws and a small mirror upon which they poured a small amount of powdered sugar. As nervous laughter filled the room, servers then placed small bottles of Agwa, a South American herbal liqueur made from coca leaves, ginseng, and guarana, in front of each person. The hosts enthusiastically encouraged diners to emulate snorting cocaine, followed by a 'chaser' of the Agwa liqueur.

Finally, the dinner concluded with a dessert course. The hosts regaled the audience with a wide-ranging conversation about morality that included drinking blood, eating fugu, the toxic Japanese sushi delicacy, and an episode from the American television show *CSI: Crime Scene Investigation* that portrayed people eating sushi off the body of a naked woman. At that particular moment in the discussion, a scantily clad young woman wearing a black leather bustier and black leather shorts strutted through the room, passing out bowls of white chocolate and strawberries. Guests scarcely had time to eat more than a few bites of dessert before the event abruptly and rather unceremoniously came to an end and event staff quickly ushered everyone out of the room.[8]

Such transgressions and restructurings are at the heart of the contemporary movement of disruptive food practices, a field that takes inspiration from an eclectic community of scientists, artists, chefs, activists, and enthusiasts who are committed to unmaking food as an ontological object. Whether disruptive food practices take place in a restaurant, exhibition space, kitchen, laboratory, garage, or even an informal hacker or maker space, disruptive food practices are oriented towards breaking rules in the pursuit of changing conversations about and experiences of food. Above all, these restructurings are philosophical and ethical projects of asking *what happens if* as a mode of radically reorganizing, and even up-ending, prevailing social systems.

One of the inspirations for this movement comes from Modernist Cuisine, a food movement with various roots, including innovation communities on the West Coast of the United States. Billing itself as a 'revolution…in the culinary arts' (Modernist Cuisine, n.d.), the Cooking Lab and Modernist Cuisine movement created by former Microsoft chief technology officer Nathan Myrhvold, brought together a community of scientists, technology professionals, artists, and chefs who were interested in the overlap between science and art, and motivated by ideals of creativity, innovation, and rule-breaking. With an academic background in theoretical and mathematical physics and a professional background in software development, Myrhvold merged his long-standing interests in scientific processes, photography, and food to create a new field of food science. Although Modernist Cuisine's use of digital technology is most immediately apparent in the movement's emphasis on innovative food photography, theoretical and practical principles from other digital realms, most notably software development, are also present. Yet above all, it is the ethos of disruption and experimentation that is associated with the digital world that is perhaps most significant.

Underscoring this emphasis on disruption as a practical philosophy for imagining and making a new world of food, are ten principles for disruption promoted by Modernist Cuisine:

1   Cuisine is a creative art in which the chef and diner are in dialogue. Food is the primary medium for this dialogue, but all sensory aspects of the dining experience contribute to it.
2   Culinary rules, conventions, and traditions must be understood, but they should not be allowed to hinder the development of creative new dishes.
3   Creatively breaking culinary rules and traditions is a powerful way to engage diners and make them think about the dining experience.
4   Diners have expectations—some explicit, some implicit—of what sort of food is possible. Surprising them with food that defies their expectations is another way to engage them intellectually. This includes putting familiar flavors in unfamiliar forms or the converse.

(Ibid.)

What characterizes these and the remaining six principles is an emphasis on the necessity and value of breaking rules – despite the irony of creating rules for breaking rules (ibid.). Ultimately, the Modernist Cuisine movement privileges play, creativity, and pleasure. At the same time, Modernist Cuisine is also focused on challenging both biophysical and intellectual boundaries: is this art or is it science?

The disruptive, transgressive, and restructuring dimensions of Modernist Cuisine and bioart appear in multiple sectors, ranging from molecular gastronomy

(Opazo, 2016; Roosth, 2013; This ,2006), to food-based fabrics, food-based instruments and soundscapes, and even coffins made out of food products. More recently, innovation industries such as those based in Silicon Valley and other technology centres have drawn on transgression to build a new food industry based on disruptive food technologies. Although established food corporations are participating in these efforts, much of the energy is coming from start-ups.

One such start-up is Reimagine Food, a global consortium with Silicon Valley roots that draws on the strengths and philosophies of the tech world to rethink food. Computer scientists, engineers, designers, artists, chefs, foodies, and many others come together to play and experiment, with aspirations of finding new opportunities to improve the global food system and ensure equitable access to food. In many ways, their projects combine science, art, agriculture, and social justice. One of their projects, termed 'Digital Gastronomy', seeks to unite the opportunities of technology and food services to create new opportunities for chefs to transform the restaurant experience. The start-up's website describes their Digital Gastronomy project as:

> The first global platform where chefs from around the world can promote their ideas, challenges and innovations while forging a closer relationship with local food producers and food and beverage companies.
>
> Through Digital Gastronomy, chefs connect their challenges and future innovations with the general public. This way, they can raise awareness of their approach and the possibilities for the future of cooking and bring them to fruition with entrepreneurs and prosumers.

The potential for revolutionizing the food system through disruption is envisioned as helping both society as a whole and individual consumers (or 'prosumers', as they are often called). Yet despite the promises of harnessing technology for disruption and revolution in the food system, fully egalitarian participation often cannot be realized. Even as technology has become more affordable and accessible, it is still restricted to those with sufficient income, facility with technology, or even access to electricity or the Internet. As Ana Davies has observed, the idea that technology will 'fix' problems is misguided; there remains a need to go beyond merely technological solutions in order to address bigger issues of 'highly unequal and unsustainable practices of eating and related food waste around the globe in the future' (2014: 182). Consequently, as critics have suggested, as long as disruptive food technologies actually instantiate difference and inequalities, they cannot be truly engaged in social justice concerns with equity. What is needed instead are projects that do not merely attempt to change the food system but rather to transform it so that it is a truly participatory and equitable food system.

Working alongside these more organized and even institutionalized forms of disruptive food technologies is another community of food disruptors, who are explicitly engaged with participatory principles and practices. Taking inspiration from design principles, participatory citizenship, and even anarchy, these activist-participants identify themselves by numerous names: citizen-scientists, democratization activists, designers, DIY-ers, makers, and even hackers. Despite their diverse backgrounds and typically a loose sense of spontaneous organization, they share a commitment to making food, science, art, activism, and even interpretation available to everyone. In so doing, their approach to participatory engagement not only changes the types of social justice projects they pursue, but also how the very questions and issues they take on are identified and understood.

Re-making the food system through hacking and DIYbio

In fall 2013, I was invited by a colleague to join a food hacking event in Prague.[9] When I asked how to prepare, my colleague simply advised me to come with an idea about something that I wanted to work on. Unsure what this meant, I came with questions but no clear sense of a project or any ideas to present in any kind of formal way. When I arrived at the rather ramshackle bar/restaurant where I was to meet my fellow participants, I watched in fascination as other people unpacked a seemingly random assortment of items from their bags: rolls of aluminium foil, electrical tape, chemistry lab measuring tools, rubber pipes, hot plates, glass flasks, packs of spice mixes, sodium alginate, and other chemistry supplies. Over the course of several days, the system of the hacking event became clear. As participants talked with one another, often over glasses of beer, wine, or cognac, they discussed projects they had wanted to try out but had not yet been able to do on their own. As people shared ideas, they encouraged one another, made suggestions on how to proceed, initiated online searches to look for suggestions, and then began taking up pieces of equipment – or even leaving in search of necessary items at a nearby grocery store, pharmacy, or hardware store – to start building tools for their experiments.

One of the projects entailed trying to turn beer into a vapour and then sampling the vapours to see how the experience differed from drinking beer in its normal beverage state. Another consisted of building a microscope from basic electronic parts culled from cameras, a smartphone, and other gadgets. Yet another project consisted of trying to reverse engineer the spice profile for a seasoning packet. Although the spice packet project is a common one for scientists working in formal, commercial laboratories and using sophisticated million-dollar instruments, the participants in our hacking workshop figured out a far less sophisticated way to use coffee filters and a ballpoint pen to do basic spectral analysis (Kera *et al.*, 2015). Although the results were not as complicated as those possible in commercial laboratories, the basic process was accomplished at a fraction of the expense (see also Delgado, 2013).

The primary goal of this food hackathon was to play, most notably to explore the possibilities of potentiality as participants asked themselves and each other questions about 'what if?' But at another level, food hackers such as these individuals are also very much concerned with addressing more familiar social justice issues of equity, safety, and access to knowledge. Play and experimentation are the methods by which more serious issues can be approached. By disrupting our expectations about food – by hacking the food system – makers want to transform food and food experiences in ways that make them accessible to as many people as possible, especially people who might otherwise be excluded from the food system as it currently exists – for instance, residents of regions with limited access to electricity or Internet service, low-income individuals, and persons without any experience or knowledge of computer programming, engineering, design, or scientific technologies more generally.

Whether labelled as food hacking, biohacking, making, tweaking, tinkering, do-it-yourself, or even we-do-it-together, these activities are oriented at creating alternative social relations, narratives, and possibilities. In describing one of their food hacking projects, Denisa Kera and Nur Liyana Sulaiman write that this project could bring about 'new forms of commensality [that] can reverse the negative effects of alienation and fast food culture' (2014: 195). Ana Delgado's account of an 'Open Source Food Night' food hacking event in Copenhagen demonstrates similar approaches: the focus of the event was on opening up and facilitating public conversations about asylum seekers in Denmark, and participants engaged in open source and open access activities to forage and create new recipes, including creating a 'hacked' yogurt (Delgado, 2013). In describing the Growbot Garden project featured as part of a series of speculative design and DIY efforts at the San Jose (California) Museum of Science and Technology in the heart of Silicon Valley, Carl DiSalvo observes that the project's emphasis on using 'robotics and sending technologies for small-scale agriculture' (2014: 237) encouraged new ideas about the nature and role of science museums and how different publics engage with science and technology (ibid.: 238). In two very different food hacking initiatives, Markéta Dolejšová has documented, on the one hand, how dieters using the Soylent food replacement have formed DIY dieting communities to trade tips, 'hack' their diets, and offer new models of healthy eating (2016a), and, on the other, how Singapore residents have used open platform technology to tweak the 'Smart Nation' rubric to reclaim ethics and practices of food sustainability (2016b). In both instances, ordinary citizens are wresting authority and expertise over knowledge, health lifestyles, and sustainability from 'official' systems as states and regulatory agencies.[10]

Central to the hacking ethos is the idea of the participatory, intellectual commons: open source, open access, open hardware, open software (Crossan *et al.*, 2016; Delgado, 2013; Kera and Suleiman, 2014). This ideal is embedded in the anarchist or design proposition that no one person owns ideas or knowledge or solutions. Instead, it is through cooperative efforts of mutual support and inspiration that ideas are generated and can then be taken up by different people who add their own ideas and possibilities (Nafus and anderson, 2009). One result of this process is that there is no predetermined direction for identifying problems or solutions. Rather, it is in the process of asking questions, playing, cooperating, challenging, and reimagining, that the horizon of possibility is opened up. Futures can never be anticipated but rather unfold through a process of emergence (Crossan *et al.*, 2016; Delgado, 2013). This is what design anthropologist Joachim Halse has described as 'moments of becoming' (2013: 194). Or, as

Caroline Gatt and Tim Ingold have argued, this mode of creating is one of improvisation, in which we '[follow] the ways of the world as they unfold ... [and as such, it] calls for both flexibility and foresight' (2013: 145). This is about 'opening up pathways' (ibid.: 145) that reveal different, and often unexpected, questions and insights.

Just as the projects that food hackers tackle are never predetermined but emerge through improvisation and experimentation, so, too, are the communities that participate in these projects. These are communities 'that [make themselves] through hacking practices of sharing, circulation, and the constant transformation of things' (Delgado, 2013: 66). As such, hackers and their participatory communities create new forms of political participation and action. In their analysis of community gardens in Scotland, Crossan *et al.* have argued that unlike 'neoliberal constructions of citizenship' that perpetuate an 'atomized citizen subject independent of any broader social responsibility or embeddedness', the DIY aspects of community gardens facilitate 'a form of political participation through a process of *learning* by being in the presence of *difference*' (2016: 937, 943). The participatory mode of DIY, making, and hacking is an emancipatory form of citizenship (ibid. 938) that is ultimately more democratic (Delgado, 2013: 67).

Food justice unmade and remade

Technological innovations such as those present in social media, modernist cuisine, bioart, and food hacking provoke important questions about the limits and possibilities of food. Collectively and individually, proponents of and participants in these different approaches raise significant theoretical and practical questions about what happens to food when it is not just decentred but denatured.

At the same time, each of these approaches to technology has transformed the nature of food justice. Issues of unfair and discriminatory labour practices and inhumane animal welfare can be addressed by using robots, drones, and synthetically produced or lab-cultured foods. Yet even as those same technological innovations can protect workers and animals, they also displace them from the food system, potentially further disenfranchising them. Three dimensional printing can offer home cooks opportunities to enjoy otherwise 'elite' meals outside expensive restaurants, even as it standardizes food by eliminating the ordinary variation that happens normally through human engagement. And the ever-present use of smartphones and social media to post pictures and reviews of food and meals can offer transparency and equity through real-time experiences that can be shared by multiple audiences, while simultaneously threatening to alienate diners from the foods in front of them and their companions at the table and flattening the fully sensory experience of eating and drinking solely to visual representations. And the use of apps and social media to encourage healthy eating and exercise also creates systems of surveillance and shaming. Thus, the privileging of technology as a means to pursue food justice issues creates both advantages and disadvantages.

Hacking, however, offers a different path through these food justice issues. Because issues and solutions are not predetermined in advance but emerge through participatory engagement, groups can identify their own concerns and set their own agendas. As a result, the diversity that is intrinsic to the hacking ethos sidesteps criticisms that food justice issues too often reflect the concerns of those who have the power to articulate priorities and acquire universality rather than reflect culturally relative situations or external, environmental factors (Biltekoff, 2013; Guthman, 2007). Additionally, the hacking ethos of disruption, unmaking, and remaking is applied broadly, not just to food or justice, but to technology itself. By asking questions and playing with technology to unmake and remake it, hackers displace the primacy of technology as the means to solve problems. And ultimately, the futurist orientation of hacking imagines multiple possibilities without needing to privilege any one as the best way forward, thereby responding to the normative ideologies that often imbue food justice efforts. Efficacy is not predetermined but rather emerges through the process itself. Consequently, hacking operates more comfortably in the space of potentiality and not-yet-known.

By not predetermining the problems to be fixed or the solutions to be used, food hackers decentre both food and technology. While individual food hackers may have their own concerns about food safety, health, well-being, sustainability, labour, animal welfare, or any other number of critical food justice issues, their commitment to collaboration and propositional modes of creation enables them to imagine multiple possibilities without needing to value any one over another from the outset. As such, it is a more democratic and liberatory mode that allows for equality of possibility.

By upsetting, transgressing, unmaking, and remaking, hackers are creating new spaces and opportunities for food justice concerns in ways that move food science, digital technology, and activism in new directions. In many respects, these disruptors are not so much innovating or implementing new forms of digital technology as they are challenging the conventions by which the digital world, as well as science and activism, are understood. By asking and practising practical philosophies of 'what if?', they simultaneously manipulate and de-privilege digital technologies. What 'the digital' offers is not necessarily new techniques or fora but rather new ways of asking questions, generating insights, and forging communities.

Notes

1 This is an on-going research project that involves both more traditional modes of ethnographic participant-observation, more experimental, design-influenced practices of participatory collaboration, and background case study research-based publicly available documents and scholarly reports. As in keeping with much work on and in speculative design, ethnographic 'sites' are fluid and spontaneous, especially as participants circulate through them and shift positions as subjects and fellow scholarly observer-analysts.

2 For instance, see the image of the farmer with a laptop on the website for the DeLaval MidiLine™ milking parlour at: www.delaval.ca/-/Product-Information1/Milking/Systems/Batch/DeLaval-MidiLine-ML2100-and-ML3100-milking-systems/ (accessed 28 April 2017).

3 In response to recurring criticisms of the contemporary industrialized food system, McClements *et al.* have acknowledged the problems caused by the food industry, especially foods that are less nutritious, while maintaining that the food industry 'has also greatly contributed to the quality of modern life' (2011: 76).

4 I am not arguing that these are new developments. Rather, noting that this distancing continues, albeit perhaps in different forms. I would, however, take issue with claims that all innovations in the contemporary food system that produce distancing are negative, as some observers would suggest.

5 For a fascinating and more detailed discussion of the transformations in productive and reproductive labours of cows and dairy farmers made possible through digital technologies, see Katy Overstreet's dissertation on Wisconsin dairy farmers.

6 While these dinners were innovative at the time, not just in Russia but elsewhere, they have since become ubiquitous as food experiences, including as 'conversation starters' at corporate events and professional conferences alike.

7 The hosts failed to make any mention of the need for food aid within Russia, long-standing practices of foreign governments and organizations providing humanitarian food aid to Russia over the past century, the fact that African communities provide food assistance and aid workers to Russia, or even any of the many Russian populist critiques of Russian assistance to Africans. For more elaboration on these issues, see Caldwell (2004).

8 Because of the event's structure, especially the rushed ending, I was unable to talk with the organizers or presenters and learn about their goals for the events.

9 I am grateful to Denisa Kera for inviting me to join the festivities and to our fellow participants at the event.

10 These are just a few of the very many examples of food hacking projects that have emerged over the past decade. Many of them take place in, or are facilitated through, university settings or science and technology museums, but they are not exclusive to those spaces. See, for instance, Banu (2014); Kuznetsov *et al.* (2016a); Kuznetsov *et al.* (2016b); Tang (n.d.); Zoran and Coelho (2011).

References

Banu, L. S. (2014) Black noise: Design lessons from roasted green chils, udon noodles, and pound cake. *The Journal of Speculative Philosophy*, 28(1): 17–30.

Biltekoff, C. (2013) *Eating Right in America: The Cultural Politics of Food and Health.* Durham, NC: Duke University Press.

Caldwell, M. L. (2004) *Not by Bread Alone: Social Support in the New Russia.* Berkeley, CA: University of California Press.

Carney, M. A (2015) *The Unending Hunger: Tracing Women and Food Insecurity across Borders.* Berkeley, CA: University of California Press.

Carruth, A. (2013) Culturing food: Bioart and in vitro meat. *parallax*, 19(1): 88–100.

Choi, J. H. and Graham, M. (2014) Urban food futures: ICTs and opportunities. *Futures*, 62: 151–154.

Contois, E. (2017) 'Lose like a man': Gender and the constraints of self-making in Weight-Watchers online. *Gastronomica*, 17(1): 33–43.

Crossan, J., Cumbers, A., McMaster, R. and Shaw, D. (2016) Contesting neoliberal urbanism in Glasgow's community gardens: the practice of DIY citizenship. *Antipode*, 48(4): 937–955.

Davies, A. R. (2014) Co-creating sustainable eating futures: Technology, ICT, and citizen-consumer ambivalence. *Futures*, 62: 181–193.

Delgado, Ana (2013) DIYbio: Making things and making futures. *Futures*, 48: 65–73.

DiSalvo, Carl (2014) The Growbot Garden Project as DIY speculation through design. In M. Ratto and M. Boler (eds) *DIY Citizenship*. Cambridge, MA: MIT Press, pp. 237–247.

Dolejšová, M. (2016a) Deciphering a meal through open source standards: Soylent and the rise of diet hackers. *CHI 16 Extended Abstracts*, 7–12 May: 436–447.

Dolejšová, Markéta (2016b) Squat & grow: Designing smart human-food interactions in Singapore. *Proceedings of SEACHI, 2016*, May 8, 2016: 24–27.

Gatt, C. and Ingold, T. (2013) From description to correspondence: Anthropology in real time. In W. Gunn, T. Otto. and R. C. Smith (eds) *Design Anthropology: Theory and Practice*. London: Bloomsbury, pp. 139–158.

Gottlieb, D. (2015) Dirty, authentic…delicious': Yelp, Mexican restaurants, and the appetites of Philadelphia's new middle class. *Gastronomica*, 15(2): 39–48.

Guthman, J. (2007) Can't stomach it: How Michael Pollan *et al.* made me want to eat Cheetos. *Gastronomica*, 7(3): 75–79.

Guthman, J. (2011) *Weighing In: Obesity, Food Justice, and the Limits of Capitalism*. Berkeley, CA: University of California Press.

Halse, J. (2013) Ethnographies of the possible. In W. Gunn, T. Otto. and R. C. Smith (eds) *Design Anthropology: Theory and Practice*. London: Bloomsbury, pp. 180–196.

Kao, C. (2016) The post-work ethic: Startup culture and emerging epistemologies of value in Silicon Valley. Paper presented at the American Anthropological Association meetings, November 2016.

Kera, D., Denfeld, Z. and Kramer, C. (2015) Food hackers: Political and metaphysical gastronomes in the hackerspaces. *Gastronomica*, 15(2): 49–56.

Kera, D. and Sulaiman, N. L. (2014) FridgeMatch: Design probe into the future of urban commensality. *Futures*, 62: 194–201.

Kuznetsov, S., Santana, C. J. and Long, E. (2016a) Everyday food science as a design space for community literacy and habitual sustainability. *CHI 16*, 7–12 May: 1786–1797.

Kuznetsov, S., Santana, C. J., Long, E., Comber, R. and DiSalvo, C. (2016b) The art of everyday food science: Foraging for design opportunities. *CHI 16 Extended Abstracts*, 7–12 May: 3516–3523.

Lyon, S. (2014) The good guide to 'good' coffee. *Gastronomica*, 14(4): 60–68.

McBride, A. E. (2010) Food porn. *Gastronomica*, 10(1): 38–46.

McClements, D. J., Vega, C., McBride, A. E. and Decker, E. A. (2011) In defense of food science. *Gastronomica*, 11(2): 76–84.

Modernist Cuisine (n.d.) modernistcuisine.com (accessed 28 April 2017).

Nafus, D., and anderson, k. (2009) Writing on walls: The materiality of social memory in corporate research. In M. Cefkin (ed.) *Ethnography and the Corporate Encounter: Reflections on Research in and of Corporations*. New York: Berghahn Books, pp. 137–157.

Opazo, M. P. (2016) *Appetite for Innovation: Creativity & Change at elBulli*. New York: Columbia University Press.

Overstreet, K. (n.d.) How to taste like a cow: Cultivating shared sense in Wisconsin dairy worlds. Unpublished manuscript.

Pennell, M. (2016) More than food porn: Twitter, transparency, and food systems. *Gastronomica*, 16(4): 33–43.

Roosth, S. (2013) Of foams and formalisms: Scientific expertise and craft practice in molecular gastronomy. *American Anthropologist*, 115(1): 4–16.

Rousseau, S. (2012) *Food and Social Media: You Are What You Tweet.* Lanham, MD: AltaMira Press.

Sexton, A. (2016) Alternative proteins and the (non)stuff of 'meat'. *Gastronomica*, 16(3): 66–78.

Spiegel, A. (2014) These are all the foods you can get delivered by drones. Huffington Post, 9 June. Available at: www.huffingtonpost.com/2014/06/09/food-delivery-drone_n_5461689.html (accessed 27 April 2017).

Stock, J. and Würger, T. (2014) Digital dairy: Robotic milk production takes over. *Spiegel Online*, 21 February 2014. Available at: www.spiegel.de/international/zeitgeist/dairy-farming-in-europe-effiency-paramount-as-eu-quotas-expire-a-954750.html (accessed 28 April 2017).

Sun, J., Peng, Z., Yan, L., Fuh, J. Y. H. and Soon,, G. (2015) 3D food printing: An innovative way of mass customization in food fabrication. *International Journal of Bioprinting*, 1(1): 27–38.

Tang, J. (n.d.) Future food hack. Unpublished manuscript.

This, H. (2006) Food for tomorrow? How the scientific discipline of molecular gastronomy could change the way we eat. *EMBO Reports*, 7(11): 1062–1066.

Wattles, J. (2017) Starbucks: Nearly a third of sales were made digitally last year. *CNN Money*, 27 April. Available at: http://money.cnn.com/2017/04/27/news/companies/starbucks-digital-sales/index.html. (accessed 27 April 2017).

Zoran, A. and Coelho, M. (2011) Cornucopia: The concept of digital gastronomy. *Leonardo*, 44(5): 425–431.

3 Diabetes on Twitter

Influence, activism and what we can learn from all the food jokes

Amy K. McLennan, Stanley Ulijaszek and Mariano Beguerisse-Díaz

Introduction

Governments across the world are increasingly using social media platforms like Twitter to disseminate health information and advice. Growing numbers of health departments and organisations have social media policies, and social media are now used by many as a low-cost tool for addressing so-called 'lifestyle diseases' such as obesity and Type 2 diabetes. The effectiveness of these initiatives is generally measured in terms of the number of subscribers following social media accounts (e.g. see Public Health England, 2014). Government social media policies tend to focus on legal concerns such as regulating staff use and ensuring privacy protection, rather than citizen health outcomes or experiences (Fast *et al.*, 2015).

Government social media policies seldom explicitly acknowledge that, for citizens, Twitter contains a multitude of messages, with public health messages being posted and read alongside marketing for unhealthy products (Kelly *et al.*, 2015); or that it may be used for a wide range of reasons, including for information seeking or dissemination (Scanfeld *et al.*, 2010), stigmatisation and exclusion, or as a source of emotional support and community acceptance (Pew Research Center, 2013). Twitter is also used as a venue for community activism and protests (Beguerisse-Díaz *et al.*, 2014; González-Bailón *et al.*, 2011), as a platform where information (or disinformation) can be shared, and activism-related events organised and coordinated. Government policies rarely consider the experience of social media users and the multitude of ways messages on the platform can be used, read, understood or interpreted; instead, they typically emphasise the importance of self-responsibility (Fast *et al.*, 2015), implying by default that it is citizens' own responsibility to safely navigate the broader social media landscape for themselves.

Content from social media that appears 'unrelated' to health advice is usually discarded by health researchers as irrelevant to their work. Analysis carried out for studies relating to Twitter and health tends to filter out content such as chatter or jokes – often about food – that the researchers view as 'noise' and 'irrelevant' to health research (e.g. Harris *et al.*, 2013; Hawn, 2009; Paul and Dredze, 2011). However, there is ample evidence that marketing (Kelly *et al.*, 2015), social values (McLennan and Ulijaszek, 2015), emotional connection, community (Ferzacca, 2004) and humour can all contribute to health outcomes by affecting lifestyle patterns, food choices, stress levels and other physiological changes (Hayashi *et al.*, 2003; McCreaddie and Wiggins, 2008).

Unlike typical policy approaches or public health studies, our multi-disciplinary approach to understanding the significance of Twitter in public health is not limited to formal health messaging or content that aligns with broader public health aims. Instead, we take a large collection of tweets containing a health term (in this case, 'diabetes'), and employ techniques from network science and information retrieval to determine who are the most influential Twitter users, what are the most common messages, and what is the content that attracts the most attention. We interrogate the significance of patterns in the data using social theory and analysis, and consider how findings fit into the broader public health context. Our collaborative analytic approach, which combines network analysis methods and social theory, is starting to break new ground in using data science to offer insights into our social world (Cihon and Yasseri, 2016).

In this chapter, we focus especially on tweets that link food, eating and diabetes, and consider what our results can tell us about the use of Twitter as a platform for food activism. While diabetes management is almost inevitably about diet and nutrition, and diabetes rates worldwide are widely acknowledged to be associated with increasing consumption of 'junk' food, diabetes-related activism on Twitter rarely directly mentions food. A number of the most seemingly authoritative accounts belong to activists and advocacy groups who use the platform to collectively organise around diabetes advocacy events and campaigns; these largely focus on raising diabetes awareness in the general population, and advocating for greater investment in research. Yet while food is surprisingly absent from tweets relating to diabetes advocacy (bringing about social or political change from *within* a system (Martinsson, 2011)) and activism (bringing about social or political change from *outside* a system (ibid.)), it is not absent from our dataset. Many users talk about food, expressing themselves in updates and jokes about what they are eating. When we look more closely at these messages, we see that some users – in this case, people who tweet about diabetes – perform consumer activism differently to diabetes advocates and activists. As a result, our work calls into question current notions of what consumer-level food activism is and does.

We begin this chapter by introducing Twitter and explaining why its content relating to diabetes and user interactions is the focus for our research. We then ask three questions. First, *whose content has the greatest influence?* To answer this question we examine the structure of the time-changing network of retweets, and introduce the notion of 'hub' and 'authority' scores from network mathematics to describe different types of influence and impact. Second, *what is the most common content?* Here, we examine the tweets generated by the top authority accounts, and explore the type of messages they disseminate. Food-related jokes and banter stand out as particularly common content. As a result, we ask *why does so much of the content involve food jokes?* We look more closely at the content that appears persistently – food-related jokes and banter – and explore what it tells us about health information seeking, community building and activism. As tweets relating to diabetes activism rarely mention food (even though diabetes is fundamentally related to food), this chapter is not directly about digital food activism. Instead, it calls into question what 'counts' as activism on social media, and highlights different forms that user-generated digital food activism might take.

Twitter is more than a news platform for people with diabetes

Twitter is a social media platform with approximately 320 million monthly active users around the world (Desilver, 2016). Users can post short public messages or *tweets* of up to 140 characters in length. In addition, users can subscribe (*follow*) to receive the tweets of other users. When users log in, they can see in their *timeline* tweets that have been posted by the people they follow, plus some sponsored tweets, which are algorithmically selected by Twitter and ordered chronologically with the most recent tweets on top. According to Twitter, the microblogging platform is 'like being delivered a newspaper whose headlines you'll always find interesting' (Twitter Inc., 2016). Many users use Twitter exclusively to read messages: over 25 per cent of users have never posted a tweet.

Twitter's marketing material emphasises the platform's importance as a global, real-time public information source; providing information about health management and diet is no exception. Many users of social media perceive health as something that can be modified by encouraging individuals to change their behaviour through education, marketing or messaging. Commercial social media platforms have a large, global user base, significant reach into people's everyday lives, and require very little infrastructure to post messages. As such, they are perceived by users as relatively cheap channels through which to encourage large numbers of individuals to change their behaviour or attitudes. Therefore, it is no surprise that Twitter and other social media have been taken up as platforms through which to seek, share and disseminate advice about a range of diet and health-related conditions that are linked strongly to so-called 'individual choice'; including diabetes, which is the focus of this chapter.

Diabetes is a growing global phenomenon (International Diabetes Federation, 2015). It is a chronic clinical condition associated with blood sugar regulation by the hormone insulin and other endocrine factors (World Health Organization, 2010). Type 2 diabetes (T2D), in which the pancreas produces insufficient insulin, accounts for around 90 per cent of cases worldwide. Type 1 diabetes (T1D) accounts for the remaining 10 per cent of cases, in which the pancreas produces no insulin at all (International Diabetes Federation, 2015; World Health Organization, 2010).

Insulin regulates a person's blood sugar levels in response to food consumption or avoidance. Foods that are high in sugar, or which have a high glycaemic index (e.g. foods high in carbohydrates and low in fibre) can cause significant blood sugar fluctuations as the sugars are rapidly absorbed and metabolised, if insulin does not effectively act as a buffer to maintain a constant blood sugar level. If a person's blood sugar becomes too low, their brain will cease to function. Over a person's lifetime, fluctuating and/or consistently high blood sugar levels can cause problems with the heart, blood vessels, eyes, kidneys and nerves (World Health Organization, 2010). People who have diabetes must constantly monitor and adjust their blood sugar levels; administering insulin or being active can reduce blood sugars, while consuming most types of food and drink can raise them.

Americans living with chronic diseases use the Internet more for communication than for seeking health information (Fox and Purcell, 2010). While they are less likely than those without chronic disease to have access to the Internet or to seek health information online, once they are online, people living with chronic disease are more likely than others to access user-generated health content such as blog posts, hospital reviews, doctor reviews, and podcasts (ibid.). If people seek emotional support when dealing with a health issue, or a quick remedy for an everyday health issue, they more commonly seek information from fellow patients, friends and family than from professional sources (Pew Research Center, 2013). Caregivers are also more likely than non-caregivers to seek health information and support online (72 per cent versus 50 per cent) (ibid.).

Twitter is increasingly being used as a platform through which to disseminate dietary and health information to people who have been diagnosed with diabetes, or who are considered to be at risk of developing it. Yet in practice, Twitter, like other social media platforms, is more than a news source that facilitates a unidirectional flow of information. It is also a venue for engaging in social interaction, seeking or giving emotional support, stigmatising or ostracising others, and organising collective action (Kwak *et al.*, 2010).

Social media such as Twitter allow users to build community through their interactions. Twitter users can interact with each other in a variety of ways, including *liking* (expressing approval or appreciation of a tweet), *replying* to a tweet, *mentioning* a specific user in a tweet, and *retweeting* (forwarding a tweet posted by someone else to one's own followers). Through retweeting, the flow of messages can create local or global concentrations of a particular type of information at a particular moment in time. For example, revelations of diabetes by celebrities such as film actor Tom Hanks spread rapidly and widely on Twitter in a short amount of time (Beguerisse-Díaz *et al.*, 2017). Twitter users generated a wealth of messages about this news; some messages were supportive while others were critical or sarcastic, some were gossip or jokes, while others mainly reported facts and news headlines. The structure of these interactions lends itself to being investigated using quantitative and computational methods, such as network science: the interactions among users (e.g. retweets, friend/follower relationships) can be represented as a network in which the users are the nodes and the relationships form the connections between them. In addition, Twitter allows its information to be mined through its application programming interface (API), which facilitates the acquisition of data in digital format.

Twitter has also been used to organise and disseminate information about collective action, from diabetes fundraising and awareness events that are explored further in this chapter, to massive protests and rallies (Beguerisse-Díaz *et al.*, 2014). Two key features contribute to making it a useful tool for advocacy and activism. From a technical perspective, it enables immediate and easy messaging so it requires few specific skills or specialised equipment. From a social perspective, it is used and endorsed by a large global user base; it cannot work as an advocacy tool unless a critical mass of users employ it in a certain way and are not prevented from doing so. Such technical and social features allow Twitter and other social media platforms to enable what is known as 'organisation without organisations' (Shirky, 2008). This new form of collective organisation permits collective action, social connectivity and consensus-formation around personal lifestyle values without group loyalties that characterised pre-digital social movements (Bennett, 2012). Individual Twitter users can, for example, pay attention to a selected group of users who match their values and interests, many of whom they may not know offline; they can choose to retweet a particular message in an instant, but they are not committed to continuing to retweet future messages about the same cause or by the same user. They can coalesce around specific issues or events momentarily, but there is not necessarily any longer dialogue or relationship formed before, during or after the digital interaction or collaboration. Other authors have described this as a distinction between 'collective action' (where formal organisations coordinate individuals in common action) and 'connective action' (where action is co-produced and shared based on personal expression) (Bennett and Segerberg, 2012).

Methods

This chapter is the result of an iterative process of collaborative research between applied mathematicians and anthropologists. The analysis is based on over 2.5 million English-language tweets that contain the term 'diabetes' posted between 26 March 2013 and 19 January 2014. The tweets were collected by Sinnia, a data analytics company.[1] Along with the text of the tweets, we collected the following information about the users who produced them:

- *Followers*: Twitter users who are subscribed to the user's tweets.
- *Friends*: Users whom the user has followed.
- *Retweets*: Tweets composed by other users that a user has passed along to his/her followers.
- *Biography*: A user's self-description, where Twitter allows users to describe themselves in 140 characters, at most.

A *network* (or graph) is an information structure (more precisely, a mathematical object) in which pairs of *nodes* (or vertices) can be connected to each other by *edges* (or arcs) (Newman, 2010). From Twitter, one can obtain several networks in which the users are nodes by creating, for example, connections that represent friend/follower relationships, or the event in which one user has retweeted someone else's tweet. In these cases, the edges are directional, that is, they distinguish the user who follows (retweets) from the user who is being followed (retweeted). The information (i.e. the content) flows in the opposite direction to the declared direction of interest. In other words, if a user 'follows' another user, the interest goes from the source of the connection to the target, and the target's tweets are received by the source. Likewise, in a retweet, the person who retweets is expressing interest in a particular message, which then is transmitted to the retweeter's followers.

We used methods from network science (analysis of centralities in temporal networks, community detection) and information retrieval (topic detection) to identify the main patterns in the content of tweets, and the interactions among the users. Importantly, only 10 per cent of user accounts in our data produced tweets that elicit any form of response (a retweet or reply). Furthermore, the intensity of the response was extremely heterogeneous: relatively few users attained a disproportionally high amount of attention. Technical sketches of our work are briefly described in the following sections.[2]

From an ethical perspective, the data presented in this study are public information (as per Twitter's terms and conditions) and do not pose ethical risks. Our analysis serves the public interest and poses no risk to users, and we do not reproduce tweets with notable amounts of sensitive or private material. Indeed, the most prominent users in our dataset also maintain other online profiles and produce tweets for public consumption. Further, users who wish to restrict access to their tweets to specific users can do so via their privacy settings. However, we do note ongoing debates about the ethical dimensions of research on social media data (Zimmer and Proferes, 2014).

Whose content has the greatest influence?

Understanding power dynamics and relationships is essential for understanding activism; activists may, for example, challenge existing powers, while empowerment can enable the success of activism. Anthropologists, sociologists, mathematicians and computer scientists, among others, have begun to explore how the rise of digital and interactive media reflects, or can change, the power landscape in global society. Some have argued in relation to food that existing power structures are destabilised and democratised as information becomes more widely available (Lien and Nerlich, 2004; Blue, 2010). In this case, attribution of 'authority' (i.e. power or rights to dominate others, rather than the technical mathematical term used in the context of a directed network) is broadened, when knowledge production becomes more participatory and widely distributed. However, others have found that a specific curatorial process is required to achieve such distributed power, and that the use of participatory platforms in the absence of careful guidelines and principles can limit their capacity to achieve the imagined ideals of open access and information democratisation (Geismar, 2012). In this case, digital platforms can just be another channel through which existing power structures are reinforced.

Quantitative approaches to investigating Twitter offer a different perspective on influence, power and authority. Using techniques from network science, researchers can find which users are more 'central' (i.e. important and/or potentially seen as important) in a network of interactions. There are many different notions of centrality that range from the simple (number of connections) to the sophisticated (using the properties of random walks on networks) (Masuda *et al.*, 2016; Newman, 2010).[3] The 'hub' and 'authority' score of a Twitter user (or node) are two examples of node centrality that are defined recursively (Kleinberg, 1999):

- A good 'authority' is a node that receives many connections from many 'hubs' (defined in the next point), i.e. authorities are users who produce tweets that attract the most attention (in the form of retweets). For example, in the worldwide web (in which web pages are nodes and hyperlinks are directed connections between them), an example of an authority would be a site such as Wikipedia, which contains content that many web users seek.
- A good 'hub' is one that points to good 'authorities', i.e. passes information from authorities to other users. In the worldwide web, an example of a hub would be Google, which contains links to other sites but which itself may not host any content.

We have extracted the 'hub' and 'authority' score of all users (a number between 0 and 1) in the weekly retweet networks in our data. In each weekly network, the sum of the authority scores of all users is equal to 1 (likewise for hub scores); the magnitude of scores can be interpreted as 'how good is this node as a hub (or authority) compared to the rest of the nodes?' (or 'how big is their share of the hub (authority) pie?').[4]

Authorities send the messages with the biggest influence

The top ten authorities in our dataset are a mixture of bloggers, advocacy groups, companies and a health information firm (Figure 3.1). The top authorities tend to have a relatively sustained presence over the data observation period (i.e. their authority score is usually not zero for most weeks). Four of the top ten authorities are directly linked to T1D. The onset of T1D is typically much earlier than T2D and tends to affect people much more severely than T2D, as it is related to the inability of the pancreas to produce insulin rather than a reduction in the pancreas's capacity to produce insulin.

We then created the follower network of the top 1,000 authorities in our data: in this network the nodes are the 1,000 top authorities, and the connections correspond to who follows whom from within this group. This network offers a different view of the relationship between these users. In a retweet network the interactions of the users is the context of a topic in a given time interval, in this case, tweets about diabetes during the data collection period. On the other hand,

Top 10 users by aggregate authority score

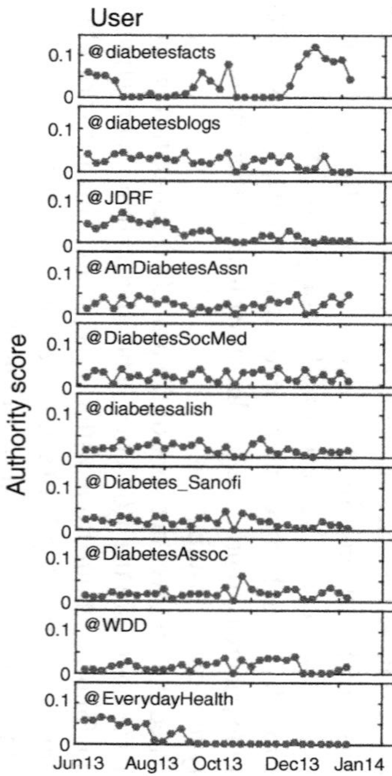

User	Rank	Aggregate score	Weeks	Description
@diabetesfacts	1	1.099528	25	News and information about diabetes from the editors of @EverydayHealth (see below).
@diabetesblogs	2	0.797916	29	Updates from Diabetes Daily, a website and blog founded in 2005 by entrepreneur David Edelman, and Elizabeth Zabell a T1D patient.
@JDRF	3	0.779938	31	Global funder of T1D research, created and led by T1D patients or people with a connection to the disease. Has strong volunteer base.
@AmDiabetesAssn	4	0.775563	31	American Diabetes Association: Advocacy and research organisation founded by a group of physicians in 1940.
@DiabetesSocMed	5	0.671353	31	Diabetes Social Media Advocacy is a programme provided by the Diabetes Community Advocacy Foundation. It was founded by Cherise Shockley (T1D patient) and obtained not-for-profit status in 2012.
@diabetesalish	6	0.637761	31	Diabetes blogger, advocate and writer, Kelly, who was diagnosed with T1D aged 8 years, and whose family also has a strong history of it.
@Diabetes_Sanofi	7	0.627676	31	Diabetes division of Sanofi, a global pharmaceutical company.
@DiabetesAssoc	8	0.586427	31	Canadian Diabetes Association, founded in 1953 to unite provincial branches.
@WDD	9	0.543949	28	World Diabetes Day (14 November) is a campaign led by the International Diabetes Federation, an umbrella organisation uniting over 230 national diabetes associations that was founded in 1950.
@EverydayHealth	10	0.524492	31	Marketing firm founded in 2002 by Ben Wolin and Mike Keriakos; it has patnerships with AOL, Google, YouTube, the Mayo Clinic, the ABC.

Figure 3.1 Top ten authorities in the data set. Top ten users by aggregate authority score, number of weeks with non-zero authority score, and brief description.

Source: Beguerisse-Díaz et al. (2017).

a follower network indicates a more 'stable' interest that need not be restricted to a specific topic: users are subscribed to receive all tweets from the users they follow, regardless of whether the tweets contain the word diabetes or not. In this follower network the users can be divided into six distinct communities using the Markov Stability community detection framework (Delvenne *et al.*, 2013). A 'community' in this context is a group of nodes that are more tightly coupled with nodes in the group than with the rest of the network (Beguerisse-Díaz *et al.*, 2014; Porter *et al.*, 2009):

C0 Health- and medicine-related accounts.
C1 A diabetes-related group of advocates, patients and families.
C2 Accounts related to lifestyle and well-being.
C3 Accounts related to news and media.
C4 Celebrities.
C5 A group of accounts specifically related to the retailer Tesco.
C6 Humour and parody accounts.

The biographies of the members of each community show a remarkable consistency in the vocabulary used (Figure 3.2); the exception is the community of humoristic and parody accounts whose members do not use the same vocabulary to describe themselves. Despite Tesco's location in the UK, it has a large presence on the global Twitter platform (in English). Aside from Tesco, food industry representatives and lobbies, which are extremely influential actors in debates

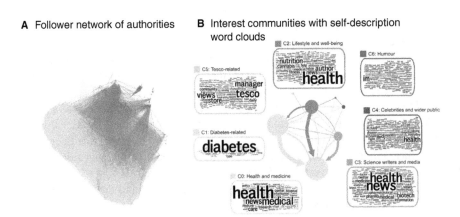

Figure 3.2 A: The follower network of the top authority account. B: The follower network coarse-grained by communities. The word clouds contain the words that most frequently appear in the members' self-descriptions. The greatest numbers of users overwhelmingly pay attention to the health and medical advice community (C0), some pay attention to the diabetes community (C1, which includes funding agencies and patients).

Source: Beguerisse-Díaz *et al.* (2017).

in relation to food policy and legislation, are notably absent from Twitter debates about diabetes.

The top authority nodes represent a variety of advocacy positions – health advocates tweet about different lifestyle choices, the well-being group promotes new diets and fads, and the pharmaceutical industry advocates for pharmaceutical intervention rather than dietary change. However, not all users give a clear signal of what underpins their position; Tesco's interest in diabetes, for instance, in addition to generating awareness, can be to promote its brand and generate opportunities to promote its products, such as health insurance and food. In practice, there is an unclear distinction between marketing, sponsorship and advocacy on Twitter. This lack of clarity is a characteristic of many electronic media: anyone can establish a presence, and there is little scrutiny of their objectives and their effects on the broader population. Users are expected to be discerning and responsible, but it is unclear on what they should base their judgement, given that many user profiles appear equally credible and the information equally 'authoritative'.

Hubs connect users to the most important messages

The top hub accounts in our dataset change from week to week, and tend to be a mixture of bloggers, automated accounts, users with no specific or declared interest in health, and accounts which have since been closed. Hubs, unlike authorities, do not have a sustained high presence over time (Figure 3.3). This means that there is no account that is routinely and consistently linking users with sources of information. Instead, hubs tend to have a flash of 'brilliance' (i.e. importance) and then dissipate. Our data do not allow any inference about why this is the case.

The top hubs in our dataset are predominantly bloggers and users who have experienced diabetes. It makes sense that they point to many authoritative sources of information, as this is their declared interest both on Twitter and on the blogs they administrate. However, the intention of the top hub – @1Medical-2News – is less clear. It appears to be a medical doctor named Dr Richard Billiard, and the account appears credible at first glance. However, this doctor has no other online presence, s/he retweets a large amount of messages per day at regular intervals (an average of 50 tweets per day since 2013), and s/he has never produced an original tweet. It is unclear who might be behind the profile or what they are attempting to achieve.

It is difficult to be a top authority and hub simultaneously

There are a number of advocates and activists in our dataset who aim to bring about social or political change. Most appear to try to do so by posting messages of their own, rather than amplifying other similar messages coming from a variety of places.

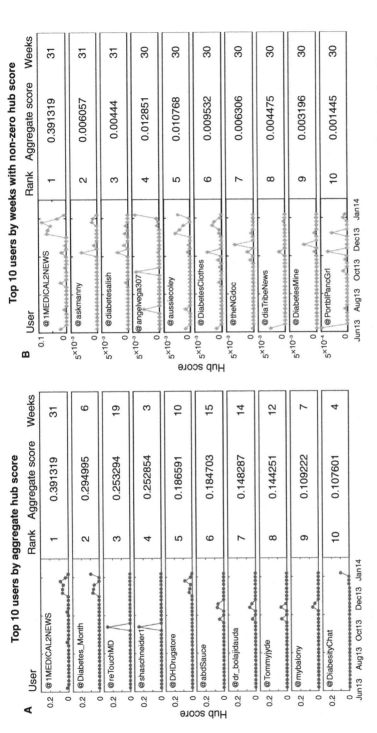

Figure 3.3 Top ten hubs in the data set. A: Top ten hub accounts by aggregate hub score. B: Top hub accounts by number of weeks with non-zero authority score.

Source: Beguerisse-Díaz et al. (2017).

Of the 2 per cent of the most central users in our dataset (by joint hub and authority scores) – users who are frequently retweeted by other users – none are at the very top of the authority and hub ranking simultaneously. Authorities tend to push out information of their own, but they retweet messages by other authoritative sources of information far less frequently. Hubs tend to retweet authorities' messages but seldom produce tweets with wide impact themselves.

There are at least three reasons why this might occur. First, authorities may all be advocating slightly different things: with no complete overlap of agendas, one organisation may not opt to pass to its followers messages from a different source. Second, it is relatively more time-consuming to read, select and retweet information from other sources; guaranteeing the reliability of every single piece of information posted by others, when the provenance cannot always be traced, can be risky for users and their reputations. This may make retweeting content an unsustainable activity, especially for established authorities that must check and verify content before retweeting it. To get around this, some users will state in their bio that a retweet 'does not mean endorsement', but this is not a realistic work-around for those running official Twitter accounts on behalf of governments or other organisations. Third, it may suggest that organisations in the public health landscape situate themselves as atomised units which together create a cacophony of messages (Lang and Rayner, 2007), rather than part of a broader network of advocates for common objectives.

What is the most common content?

In addition to knowing who the most influential users on Twitter are, it is useful to understand the content of the messages being posted. Content in tweets can be aggregated into topics using methods from information retrieval and natural language processing, and then qualitatively analysed to interrogate and explain themes and patterns (Beguerisse-Díaz et al., 2017). We employ a variant of the Latent Dirichlet Allocation method which computes the probability that a document (tweet) belongs to a topic based on the words it contains, and the words in the other documents (Blei et al., 2003). We then use an inductive grounded theory approach to manually classify the topics of the tweets into four broad thematic groups (see also Beguerisse-Díaz et al., 2017; Bowen, 2008; Braun and Clarke, 2006): health information (e.g. health advice); news (e.g. headlines about a latest breaking story, described in more detail in the following section); social interaction; and commercial. One anomalous cluster of recurrent tweets is discussed further in the following section. The word clouds of the top 200 word roots in each thematic group (e.g. the word root 'obe' stands for 'obese', 'obesity', and so on) in each theme are shown in Figure 3.4. Word clouds were created using script in Python, where the size of the word in the cloud is roughly proportional to how frequently the word appears in tweets that belong to each group.

Heath information

News

Social interaction

Commercial

Recurrent content

Most active users

Figure 3.4 Word clouds of the four thematic groups and recurrent content. The clouds in the left column are formed by the most frequently used terms in the thematic group (larger words appear more frequently). The clouds in the right column are formed by the names of the users whose tweets appear more frequently in the thematic group (users who appear more in the group appear larger).

Source: Beguerisse-Díaz *et al.* (2017).

Content can be split into four main thematic groups

Through iterative thematic coding, four thematic groups clearly emerge – health information; news; social interaction; and commercial. In the first group, health information, research findings, recommendations, advice and warnings, are abundantly tweeted and retweeted by a range of users. The top ten terms in this thematic group are: *risk, type 2, disease, heart, research, month, obesity, fruit, news* and *aware*. Tweets in this group include:

- Public health messages.
- Links to articles, blogs and studies about risks, treatment and cure of diabetes.
- Population health fears.
- Publicity about outreach and awareness events and activities.
- Advice about diabetes management and diagnosis.
- Lifestyle, diet and cookery tips, news and links.
- Life stories and experiences (some for marketing purposes).
- Dangers of sugar, sugar replacements and/or soda.

The advice in these tweets generally appears authoritative in tone and language, with confident and impersonal 'statements of fact', making it difficult to distinguish less-credible advice from more credible advice. For example:

@Achieveclinical: 8 Tips for Eating Out With Diabetes – Type 2 Diabetes Center-Everyday Health http://t.co/u5nIZ4cg5E #diabetes #health #diettips

@down2earthindia: Fighting flab? Think before u reach out 4 sugar substitute #Sugar #Sucralose #Diabetes http://t.co/cfeYURK7r3

@pinkdrinkladysr: Diabetes is a disease that can strike when you don't take care of your body. Check out these eye-opening statistics. http://t.co/zwpfTPgbtu

Credibility might be discerned from an examination of the original source of the tweet; however, user accounts are not always forthcoming with information and legitimacy is difficult to discern (Beguerisse-Díaz *et al.*, 2017). In general, there is a high turnover in the content that each user is exposed to, even though many messages (e.g. those from newspapers and online media) are posted multiple times. Put another way, a 'hot topic' in one week will not necessarily appear in subsequent weeks.

The second group contains news-related content. News tweets in the dataset list a headline of a news article and sometimes the first line of the story, and often provide a link to the complete story. For example:

@wwhitworthmw: AstraZeneca could buy Bristol stake in diabetes JV: analyst: LONDON (Reuters) – AstraZeneca may seek to increa... http://t.co/rcZQPIIZxo

@DietitianInNYC: Fish Oil Pills Might Cut Diabetes Risk, Researchers Say http://t.co/fvsIMXFKGR

@KatieBaby4587: Thief steals family car with daughters #diabetes medicine insidehttp://youtu.be/juPZtMmzL2s via @youtube @tmz http://t.co/agDquGWmUL

The top ten terms in this group are: *type 2, risk, fruit, type 1, eat, people, blueberry, cut, research* and *juice.* Some news-related tweets communicate research breakthrough studies or technologies, which may be reported with messages of hope for those who have diabetes, in particular T2D. Tweets in this group include:

* Headline links to particular 'breakthrough' studies or technologies.
* Celebrity news.
* General news articles about diabetic people or pets.
* News relating to the pharmaceutical industry and the economy.

The third group corresponds to social interaction. These tweets use language differently than the other thematic groups: they are typically informal and conversational in tone, their attention to spelling and grammar is limited, and they often use exclamation marks and other punctuation (e.g. emoticon smiley faces) to express fun, laughter, exasperation and abuse. The top ten terms in this group are: *give, health, food, die, think, fat, year, diet, disease* and *cause.* Tweets typically include:

* Users who joke about how what they have eaten is likely to give them diabetes.
* Chatter and everyday social interchanges that include mentions of diabetes.
* Everyday experiences of diabetes.
* Stigmatising comments.
* Banter, sexual innuendo and humour relating to sweetness and diabetes.

These tweets indicate a baseline level of awareness of dietary guidelines and diabetes aetiology. Users have conversations and interact about a diversity of topics in chatter that is not necessarily directly related to diabetes but may include references to it. People who have diabetes – particularly T1D – also talk about the daily experiences of their bodies, sugar management, and social acceptance or stigma; such tweets may elicit retweets or messages of support from other users. Some users also talk in terms of a division between 'us' (people with diabetes, especially T1D) and 'them' (people without diabetes). For example, a user talks about T1D as being a feature he/she looks for in a romantic partner:

I haven't stopped thinking about this girl for seriously like…a month. AND she has diabetes! #diabetesperks

Such content often receives retweets and replies, including messages of support, or appreciation of a joke.

On the other hand, stigmatising comments, especially tweets that blame diabetic people for bringing the disease on themselves through, for example, poor diet or lack of physical activity, are abundant in the dataset. Faced with such messages, users with T1D diabetes point out that it is important to differentiate between T1D and T2D, insinuating that while T1D diabetes is not a person's 'fault', T2D may be. Other tweets include calling other people 'diabetic' as an insult and wishing diabetes upon a person a user does not like.

A distinct theme in this category consists of tweets with sexual innuendo. At their mildest, such tweets refer to boy-band members or other (e.g. celebrity) infatuations, where the person is said to be 'so sweet' they are diabetes-inducing. At their most extreme, such tweets joke that others' bodily fluids and genitals are so sweet they are diabetes-inducing, and these tweets contain links to pornography websites or other explicit material. Like the jokes discussed earlier, these tweets reflect a baseline awareness of the links between sugar and diabetes among people who do not appear to have diabetes themselves.

Finally, commercial tweets advertise products, jobs and pharmaceuticals. For example,

@vernhenderson99: American-Diabetes-Wholesale: $12 Off Order of $100 or More! Code: ADW12100 http://t.co/e5K20ptIhH

@4londonjob: #jobs,#ukjobs Clinical Nurse Specialist Diabetes http://t.co/QtnCwYK6WR #jobs4u

@AmatoOrganogold: Caffeine stimulates elevated of Cortisol = arthritis, obesity, diabetes, and depression. Try healthy coffee: – http://t.co/4paSPZ0mrc

The top ten terms in this group are: *type 2, drug, job, manage, care, health, marijuana, sale, test,* and *for sale*. Common tweets include:

- Advertisements for jobs in the pharmaceutical and care industries.
- Marketing for a specific product, app, treatment, event or service.
- Pharmaceutical, health industry and stockmarket updates and FDA approvals.
- Sales of diabetes drugs, diets or treatment products online.

Diabetes treatment and management form a lucrative industry because diabetes is a chronic condition that requires regular and life-long treatment (rather than a cure), and so the demand for pharmaceutical products and lifestyle aids is inelastic (Simonsen *et al.*, 2015). People with diabetes depend on different technologies, consumables, health services and pharmaceutical products. Furthermore, the number of people with T2D is projected to increase dramatically in the future as a result of population ageing and obesity (International

Diabetes Federation, 2015), which will further expand the market. This commercial dimension of diabetes is reflected in many Twitter messages similar to the examples above.

Who contributes which content, and what do users advocate for?

Overall, tweets are posted by users with different claims to expertise: individuals who have first-hand experience of diabetes; personal trainers advertising their services; companies selling lifestyle products or services; other users with an apparent interest in diabetes, cookery and 'healthy' eating; marketing agencies trying to sell a particular food, supplement or device; or hospitals and health agencies attempting to communicate a specific health message. Home remedies and 'miracle cures' appear alongside health tips and recommendations. Other health-related messages include publicity about outreach and awareness events, activities and information.

The topics in which the highest numbers of top ten authorities converge are related to diabetes advocacy and awareness. For example, a topic about Diabetes Blog Week in May 2013 gathered six of the top ten authorities: *@diabetesalish*, *@diabetesblogs*, *@DiabetesSocMed*, *@Diabetes_Sanofi*, *@diabetesfacts*, and *@EverydayHealth*. In other weeks, the top ten authorities appear together in topics related to promotion of blogs by diabetics (using the hashtag *#dblogs*, which appears in 15,901 tweets in the dataset), and diabetes social media awareness (using the hashtag *#dsma*, which is promoted by @DiabetesSocMed and appears in 10,945 tweets).

All of the top ten authorities post messages relating to health information frequently and consistently. Some also feature news-related tweets, although these are less common. Two accounts, *@Diabetes_Sanofi* and *@diabetesblogs*, post a broad range of health information tweets. Two other accounts, *@WDD* and *@AmDiabetesAssn* primarily contribute tweets related to outreach and advocacy activities, events and news. The not-for-profit organisation and research funding body *@JDRF* produces tweets that contain life stories and experiences of diabetes sufferers more than any other top ten authority. Importantly this is not interactive or interpersonal in any way, but appears to be just a different framing of news and information. In these users' tweets, the boundaries between health information, health promotion, research and advocacy are blurred.

Two accounts, *@diabetesfacts* and *@EverydayHealth* (both owned by Everyday Health, Inc.) focus predominantly on lifestyle and diet-related tips, hints and advice. Unlike the other authorities, these do not produce the same outreach or advocacy messages in which users advocate for the rights and well-being of a group of people with diabetes. Instead, the majority of the tweets from these two accounts provide a link back to the company's website, which offers articles containing health and lifestyle advice. This illustrates the often blurred distinction between advocacy and advertising when using this digital medium.

The messages posted by the accounts *@diabetesalish*, a blogger and diabetes advocate who has had diabetes for over 30 years, and *@DiabetesSocMed*, a diabetes social media advocacy group founded by a T1 diabetic, are dominated by a mix of social interactions, banter and advocacy. They post content relating to health information and news headlines, but to a lesser degree than the other top ten authorities. Their tone is different to the others: it is informal and conversational rather than authoritative or informational. For example:

@DiabetesSocMed: Happy Mother's (aunts, fur baby moms, god moms, etc.) to all the women in the diabetes community! Have a great day!

or

Banging My Head Over Hubby's Clueless Doc http://t.co/2EfKqA2qZZ #diabetes #dblog

Two other users, @diabetesblogs and @DiabetesAssoc, also tweet some social and interpersonal messages.

Two accounts, @diabetesblogs and @diabetesalish, occasionally feature marketing or product promotion messages. This is common practice among bloggers, who often both advocate for a particular issue and generate income by advertising goods and services. Indeed, marketing on Twitter is not necessarily as straightforward as having a user profile representing a company posting advertising messages. Firms can advertise, lobby or seek to influence on the platform in much less direct ways, for example, by being posted on bloggers' sites, sponsoring organisations, events or individuals, or by posting and retweeting messages through accounts appearing to be unrelated to the company in question. This may reflect loopholes in regulatory practices in many countries: while company advertising and marketing are often regulated by state authorities, company sponsorship of bloggers is not regulated, and nor is the commercial content that bloggers post.

Why does so much of the content involve food jokes?

Humorous tweets, jokes and memes generate substantial and sustained interest over time, something that other types of tweets, and often other forms of advocacy, seldom achieve. They also mention food and drink more than any other type of tweet.

The intensity of collective activity (e.g. number of tweets, book sales, Internet searches) can follow a pattern of spikes of interest followed by a relaxation, driven by either external (exogenous) events, or internal activity (endogenous) (Sornette *et al.*, 2004). For example, tweets about Tom Hanks in our dataset began to appear rapidly after Hanks revealed he had T2D on a talk show on 8 October 2013, but interest subsequently waned (Figure 3.5, dashed line). However, several tweets in our dataset have an activity profile that is strikingly distinct from what we would expect to see; they have a high, sustained occurrence rate over a long period of time. For example, a joke about mathematics and diabetes appears consistently (Figure 3.5, dotted line). The top ten terms in this group of recurrent tweets are: *sugar, eat, blood, sweet, risk, type 2, drink, high, reduce* and *health*.

Humorous tweets not only generate sustained interest, they also maintain similar phrasing across the duration of the dataset. Examples of these tweets include:

- Jokes about the relative healthiness or unhealthiness of a particular food or activity, relative to widely published public health standards.
- Lyrics from two specific rap songs (one making a joke about sex, sugar and diabetes; the other an inspirational song by a rap artist with T1D).
- Viral 'fun facts' or trivia such as tasting urine as a test for diabetes, or moderate consumption of alcohol being linked to reduced diabetes.

One of the most prominent instances of recurrent content in our data corresponds to various versions of a mathematics joke:

Math Problems: If Jim has 50 chocolate bars, and eats 45, what does he have? Diabetes. Jim has diabetes…

This joke appears consistently in our dataset, more than any other specific tweet (44,130 times including retweets). Other common jokes are exclamations that a user's latest meal or snack (typically food products, from soda to cookies to ice cream) was tasty but will likely cause them diabetes:

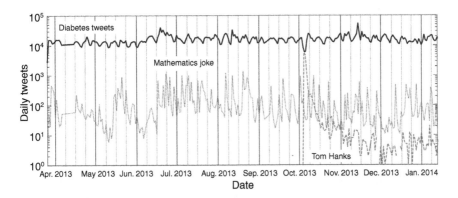

Figure 3.5 Intensity of collective activity for different types of tweet in the data set. The number of daily tweets in English containing the term 'diabetes' (solid line), the number of tweets containing some version of the 'mathematics joke' (dotted line, appears in a total of 44,130 tweets in our dataset), and the number of tweets mentioning the actor Tom Hanks, who disclosed that he had diabetes in October 2013 (dashed line, appears in a total of 13,454 tweets in our dataset).

Source: Beguerisse-Díaz *et al.* (2017).

2 bowls of yogurt, a bowl of oreos, hersheys, chips, cheese and a shitload of mints. My diet consists of diabetes.

Bother! Burger King has arrived. Hello obesity, diabetes, poor nutrition. McD's is bad enough. Grumble Grumble????

The coca cola Christmas advert, because nothing says Christmas quite like diabetes and capitalism. LOL

It would be easy to disregard these tweets as 'noise' distracting from more important messages; indeed, one anonymous reviewer of our initial manuscript that reported this finding asked why we were writing about such material (Beguerisse-Díaz *et al.*, 2017). It could also be the case that the format of Twitter (with short messages only) lends itself to short, light statements only, rather than other types of communication, although this does not necessarily explain the jokes and humour in particular.

However, there are several reasons why it is important to examine these tweets further. First, the striking consistency and volume of these messages, especially compared to other tweets in the dataset, must be important in some way. Second, just because these tweets contain content that does not fit into the narrow biomedical definition of health and nutrition, does not mean the content is unrelated to it. Indeed, these tweets represent how a large proportion of Twitter users engage in conversations about diabetes. They also discuss food – which is fundamentally related to diabetes, but not frequently mentioned in other Tweets about diabetes in our dataset. For this reason in the following pages, we look further at humour, and food jokes in particular.

Humour is not a well-studied topic in anthropology and the social sciences (Wasilewska, 2013). Some anthropologists have written about it, predominantly because jokes or humour inevitably arise in the course of fieldwork and so ethnographers talk about them briefly (Carty and Musharbash, 2008). Sustained enquiry on the topic has been much less common.

The investigations that have been carried out suggest that humour largely appears to relate to social inclusion and exclusion, where laughing *with* some people inevitably leads to others being excluded from the joke or even being laughed *at*. Digital platforms permit users to create their own categories, and jokes can illuminate what those categories are.

A number of other common themes arise in research relating to humour and jokes (ibid.). In particular:

- Laughter and humour are often used to express unease about discrimination, domination and power imbalances.
- Laughter often appears to have a role in mediating social rupture.
- Laughter and humour often only make sense in a particular context (time, place and/or social context).

These anthropological insights highlight just how significant the relatively high and enduring popularity of diabetes-related jokes on Twitter might be for understanding the success – or failure – of health messaging and advocacy.

Joking on Twitter is set in a global context, at a time where there is good public awareness about rising diabetes levels, and the links between diet and diabetes. Jokes are also set in a context of rising neoliberal forms of governance that emphasise free markets, consumerism and self-responsibility. Individuals are expected to navigate markets responsibly and to avoid doing things that might harm them, and at the same time to be consumers of products and services. We have discussed this consumer-citizen tension elsewhere (Ulijaszek and McLennan, 2016).

The majority of diabetes jokes on Twitter point to a sense of powerlessness among users in this neoliberal, globally connected context. The chocolate bar joke, for example, and the multitude of other jokes about what people have eaten that is likely to give them diabetes, make a subtle protest, a mockery of official dietary advice, as consumers at once acknowledge what they 'should' be eating, jest that they have capitulated again in the face of the ubiquity of unhealthy food, and ironically resign themselves to developing the inevitable – diabetes. Excluded by these jokes are health agencies and their scientific advice, perhaps purposefully. Included are everyday people, who demonstrate that they understand the scientific advice but also recognise the cultural dominance of the foods they invoke in their everyday lives (soft drinks, cookies, ice cream, chocolate, junk food, and so on). Medical and scientific advice is quickly shown to be almost ridiculous when placed in the context of a world in which food products entice us at every turn. Jokes in this case appear to represent critique or unease due to power imbalances.

Some jokes are specific to diabetic patients: humour helps to build camaraderie between diabetes sufferers, who bond by joking about what they have to go through on a daily basis to survive. Irony and sarcasm, in particular, are used in this case as people who have diabetes build a community through shared experience. The importance of social support and inclusion for healing has been noted elsewhere (Holt-Lunstad *et al.*, 2010; Kawachi and Berkman, 2000), although it is unclear whether this form of collective or connective action might be considered a deliberate or explicit activist practice.

Some jokes advocate for a different type of justice: a justice based on the view that individuals should each benefit from public funds and health services proportionately, and that those with chronic diseases such as diabetes 'unfairly' use significantly more resources than others. Further, when diabetes is framed as resulting from individual choices (a common view in neoliberal societies in particular), advocates for this position argue that the individual should therefore have to pay for the consequences themselves rather than be supported by society. This view is typically expressed through stigmatisation and abuse. Stigmatisation of obesity is well documented (Brewis, 2014), and although it is not as common as jokes about food and diabetes in general, there is evidence of similar stigmatisation of people with diabetes in our dataset. In these instances, people with T1D were quick to highlight that theirs was not a form of diabetes caused by being irresponsible consumer-citizens. More broadly, instances of this form of humour are used to advocate for tighter reprimands for people who are perceived to self-impose themselves as burdens on society.

Conclusion

The collaborative, multidisciplinary approach we are developing has the potential to break new ground in understanding sociocultural patterns in today's digitally enabled global society (Cihon and Yasseri, 2016). Our investigation of diabetes on Twitter clearly illustrates that this social media platform is much more than a news source. It is a site for social interaction and support. It is a site through which collective action and advocacy are organised and coordinated, especially for raising awareness about a particular cause or event. It is also a site where 'connective action' (Bennett and Segerberg, 2012) occurs. This dynamic organisation around a particular idea – such as using humour to highlight consumers' sense of powerlessness – brings to light shared values but does not necessarily connect to action in the non-virtual world.

Digital platforms are commonly assumed to make disruption of the social order possible, and to democratise knowledge and power. For example, digital platforms arguably permit users to choose where to direct their attention, and to collectively give more authority to some voices over others. However, science and technology studies (STS) and media scholars have commonly called this utopian view of social media and user agency into question (e.g. Ruppert, 2015; van Dijck, 2009). Several findings in our analysis contribute to this problematisation.

In the digital environment, it is difficult to tell where the real powers lie, especially in relation to health and well-being. Health agencies often invoke numbers of followers as an indicator of influence, but this does not consider the impact of content posted, nor the complex structure of the digital environment (e.g. the structure of the underlying networks of interest and interaction). Calculating authority and hub scores using data science represents a more sophisticated approach that acknowledges and exploits network structure. While this gives more information about users who have most influence, it does not necessarily reveal what positions the most influential users are advocating, or why.

A hub's role of bridging content between users or distributing information can be an important one, especially when it comes to seeking to elicit widespread political or social change. However, it does not appear to be a sustained activity by any one user on Twitter in relation to diabetes. It may be worth considering whether this represents an opportunity to improve public health information. This could be capitalised on if large organisations focus not only on publishing their own information, but also on pointing at relevant *and reliable* information from other sources, and engaging more widely with other Twitter users. In other words, leading authorities could become more hub-like to maximise their influence.

Overall, the users who are most influential on Twitter when it comes to diabetes are a diverse group. This challenges the notion that the domain of health is a discrete category: social media content relating to diabetes on Twitter is connected to a diverse range of sectors, organisations, interests and perspectives. Biomedicine and public health must look beyond their boundaries to identify important influencers of people, their opinions and their behaviours.

The top 1,000 authority users extracted from our analysis are similar to many of the authorities that we would expect to see based on observations in the non-virtual world, including health authorities, lifestyle coaches, pharmaceutical firms and celebrity chefs (Beguerisse-Díaz *et al.*, 2017). However, there is a notable exception: aside from retailer Tesco, the food industry is poorly represented. The absence of the food industry may be related to our use of hub and authority scores for this analysis. Both of these rely on a response from the wider public (i.e. retweets) in order to achieve a high authority score. The absence of the food industry in this case could be interpreted as users choosing to direct their attention elsewhere and to ignore organisations considered to be powerful in the non-virtual world. Superficially, this might support the claims that digital platforms like Twitter can democratise knowledge and power.

However, this interpretation assumes that influence is primarily exerted in direct and obvious ways. Looking more closely, the food industry is present in our data in other ways. Some advertising appears in the blogger profiles that we investigate, some authorities receive funding and sponsorship from companies (e.g. Tesco sponsors Diabetes UK (Tesco PLC, 2014) and soda companies sponsor health organisations (Aaron and Siegel, 2016)). Sponsorship, corporate philanthropy, advocacy and lobbying expenditures are largely undocumented and unregulated by nation states (ibid.), so the extent to which corporate powers also exert power in the landscape of social media and health is unknown.

Food brands appear frequently in many users' tweets, especially jokes about foods being likely to give them diabetes. Coca-Cola, Hershey's, McDonald's, Burger King and Oreos are mentioned, to name a few. Users' jokes about these brands imply a sense of powerlessness in the real-world food environment. Humorous tweets that tend to be considered irrelevant to public health researchers and policy-makers suggest a public resigned to the dominance of certain food products in everyday life. Users appear to jokingly set their everyday food environment against health advice about how to avoid diabetes. The humorous and flippant quips simultaneously convey a deeper ironic observation about the disjuncture between the food environment and dietary advice.

George Orwell observed that 'a thing is funny when – in some way that is not actually offensive or frightening – it upsets the established order. Every joke is a tiny revolution' (Orwell, 1945). Orwell pointed out that jokes can highlight the relative weakness of established powers, and use of humour can disrupt the established hierarchy. He also observed, that 'whatever destroys dignity, and brings down the mighty from their seats, preferably with a bump, is funny. And the bigger they fall, the bigger the joke' (ibid.). If the size of a joke is crudely measured by the number of times it is posted, or the number of retweets it obtains, then the biggest jokes on Twitter where diabetes is concerned bring down the mighty health authorities, admit powerlessness in the face of omni-present big-brand food products, and remind them that the world of food and health is much bigger than their narrow interpretations of it. Governments and researchers position themselves as authorities in this space, often downplaying or ignoring the effect of corporate lobbies, so they are the easy target of these jokes. At the same time, users ironically point to where they feel the real power lies – a power that creeps into their daily lives in myriad ways, but which demonstrates its authority in its pervasive actions rather than through statements or proclamations. Everyday citizens joke about which foods are going to give them diabetes, and in doing so, highlight an uneven balance of power between citizens, governments and organisations that advocate for healthy diets, and the world around us.

Acknowledgements

MBD acknowledges support from the James S. McDonnell Foundation Post-doctoral Program in Complexity Science/Complex Systems Fellowship Award (#220020349-CS/PD Fellow), and the Oxford-Emirates Data Science Lab. All authors would like to thank Guillermo Garduño of Sinnia for his assistance with the data collection, and Mauricio Barahona for his contribution to an earlier manuscript.

Notes

1 Sinnia is a data analytics company operating from Mexico City. More information can be found at their website: www.sinnia.com (accessed 15 October 2017).
2 For a detailed technical explanation of the mathematics and algorithms, we refer readers to Beguerisse-Díaz *et al.* (2017).
3 Random walks are useful processes to study networks. In a nutshell, a random walk consists of one 'walker' or an ensemble of 'walkers' that navigate the network. When a walker arrives at a particular node it decides where to go next by choosing one of the node's connections at random. There is a long history of using the properties of random walks to investigate the properties of systems in the life, social and physical sciences. For a review of random walks on networks, see Masuda *et al.* (2016).
4 See Beguerisse-Díaz *et al.* (2017) for a precise mathematical formulation and imple-mentation of the problem.

References

Aaron, D. G. and Siegel, M. B. (2016) Sponsorship of national health organizations by two major soda companies. *American Journal of Preventive Medicine*, 52(1): 1–11.

Beguerisse-Díaz, M., Garduño-Hernández, G., Vangelov, B., Yaliraki, S. N. and Bara-hona, M. (2014) Interest communities and flow roles in directed networks: The Twitter network of the UK riots. *Journal of the Royal Society Interface*, 11(1): 20140940.

Beguerisse-Díaz, M., McLennan, A. K., Garduño-Hernández, G., Barahona, M. and Uli-jaszek, S. J. (2017) The 'who' and 'what' of #diabetes on Twitter. *Digital Health*, 3: 1–29.

Bennett, W. L. (2012) The personalization of politics: Political identity, social media, and changing patterns of participation. *Annals of the American Academy of Political and Social Science*, 644: 20–39.

Bennett, W. L. and Segerberg, A. (2012) The logic of connective action: Digital media and the personalization of contentious politics. *Information, Communication and Society*, 15(5): 739–768.

Blei, D., Ng, A. and Jordan, M. (2003) Latent Dirichlet allocation. *Journal of Machine Learning Research*, 3: 993–1022.

Blue, G. (2010) Food, publics, science. *Public Understanding of Science*, 19(2): 147–154.

Bowen, G. A. (2008) Naturalistic inquiry and the saturation concept: a research note. *Qualitative Research*, 8(1): 137–152.

Braun, V. and Clarke, V. (2006) Using thematic analysis in psychology. *Qualitative Research in Psychology*, 3(2): 77–101.

Brewis, A. A. (2014) Stigma and the perpetuation of obesity. *Social Science and Medicine*, 118: 152–158.

Carty, J. and Musharbash, Y. (2008) You've got to be joking: asserting the analytical value of humour and laughter in contemporary anthropology. *Anthropological Forum*, 18(789296667): 209–217.

Cihon, P. and Yasseri, T. (2016) A biased review of biases in Twitter studies on political collective action. *Frontiers in Physics*, 4(34): 1–10.

Delvenne, J-C., Schaub, M. T., Yaliraki, S. N. and Barahona, M. (2013) The stability of a graph partition: A dynamics-based framework for community detection. In A. Mukherjee *et al.* (eds) *Dynamics On and Of Complex Networks*, vol. 2. New York: Springer, pp. 221–242.

Desilver, D. (2016) 5 facts about Twitter at age 10. Pew Research Center. Available at: www.pewresearch.org/fact-tank/2016/03/18/5-facts-about-twitter-at-age-10/ (accessed 13 September 2016).

Fast, I., Sørensen, K., Brand, H. and Suggs, L. S. (2015) Social media for public health: An exploratory policy analysis. *European Journal of Public Health*, 25(1): 162–166.

Ferzacca, S. (2004) Lived food and judgments of taste at a time of disease. *Medical Anthropology*, 23(1): 41–67. Available at: www.ncbi.nlm.nih.gov/pubmed/14754667 (accessed 21 July 2011).

Fox, S. and Purcell, K. (2010) Chronic disease and the internet. *Chronic disease and the internet*. Available at: www.pewinternet.org/2010/03/24/chronic-disease-and-the-internet/ (accessed 11 September 2016).

Geismar, H. (2012) Museum + digital = ? In H. A. Horst and D. Miller (eds) *Digital Anthropology*. London: Berg, pp. 266–287.

González-Bailón, S., Borge-Holthoefer, J., Rivero, A. and Moreno, Y. (2011) The dynamics of protest recruitment through an online network. *Scientific Reports*, 1: 197.

Harris, J., Mueller, N., Snider, D. and Haire-Joshu, D. (2013) Local health department use of Twitter to disseminate diabetes information, United States. *Preventing Chronic Disease*, 10(3): 120215.

Hawn, C. (2009) Take two aspirin and Tweet me in the morning: how Twitter, Facebook, and other social media are reshaping health care. *Health Affairs*, 28(2): 361–368. Available at: www.ncbi.nlm.nih.gov/pubmed/19275991 (accessed 1 November 2013).

Hayashi, K., Hayashi, T., Iwanaga, S., Kawai, K., Ishii, H., Shoji, S. and Murakami, K. (2003) Laughter lowered the increase in postprandial blood glucose. *Diabetes Care*, 26(5): 1651–1652.

Holt-Lunstad, J., Smith, T. B. and Layton, J. B. (2010) Social relationships and mortality risk: A meta-analytic review. *PLoS Medicine*, 7(7): e1000316. Available at: www.pubmedcentral.nih.gov/articlerender.fcgi?artid=2910600&tool=pmcentrez&rendertype =abstract (accessed 29 July 2011).

International Diabetes Federation (2015) *IDF Diabetes Atlas*, 7th edn. Brussels: International Diabetes Federation.

Kawachi, I. and Berkman, L. (2000) Social cohesion, social capital and health. In L. Berkman and I. Kawachi (eds) *Social Epidemiology*. Oxford: Oxford University Press, pp. 174–190.

Kelly, B., Vandevijvere, S., Freeman, B. and Jenkin, G. (2015) New media but same old tricks: Food marketing to children in the digital age. *Current Obesity Reports*, 4(1): 37–45.

Kleinberg, J. M. (1999) Authoritative sources in a hyperlinked environment. *Journal of the ACM*, 46(5): 604–632.

Kwak, H., Lee, C., Park, H. and Moon, S. (2010) What is Twitter, a social network or a news media? In *Proceedings of the 19th International Conference on World Wide Web (ACM)*, pp. 591–600.

Lang, T. and Rayner, G. (2007) Overcoming policy cacophony on obesity: An ecological public health framework for policymakers. *Obesity Reviews*, 8(S1): 165–181.

Lien, M. E. and Nerlich, B. (eds) (2004) *The Politics of Food*. London: Berg.

Martinsson, J. (2011) #2: Activism versus advocacy. *People, Spaces, Deliberation (External Affairs Operational Communication, World Bank)*. Available at: http://blogs.worldbank.org/publicsphere/activism-versus-advocacy (accessed 29 January 2017).

Masuda, N., Porter, M. A. and Lambiotte, R. (2016) Random walks and diffusion on networks. *arXiv*, 1612.03281.

McCreaddie, M. and Wiggins, S. (2008) The purpose and function of humour in health, health care and nursing: A narrative review. *Journal of Advanced Nursing*, 61(6): 584–595.

McLennan, A. K. and Ulijaszek, S. J. (2015) An anthropological insight into the Pacific Island diabetes crisis and its clinical implications. *Diabetes Management*, 5(3): 143–145.

Newman, M. (2010) *Networks: An Introduction*. Oxford: Oxford University Press.

Orwell, G. (1945) Funny, but not vulgar. *Leader*, 28 July. Available at: http://orwell.ru/library/articles/funny/english/e_funny (accessed 17 July 2016).

Paul, M. J. and Dredze, M. (2011) You are what you tweet: Analyzing Twitter for public health. *International Conference on Weblogs and Social Media (ICWSM)*.

Pew Research Center (2013) Health fact sheet. Available at: www.pewinternet.org/fact sheets/health-fact-sheet/ (accessed 11 September 2016).

Porter, M. A., Onnela, J-P. and Mucha, P. J. (2009) Communities in networks. *Notices of the American Mathematical Society*, 56: 1082.

Public Health England (2014) Public Health England Marketing Strategy: 2014 to 2017, p. 25. Available at: http://cogprints.org/280/1/getreal.htm (accessed 15 October 2017).

Ruppert, E. (2015) *Being Digital Citizens*. London: Rowman & Littlefield.

Scanfeld, D., Scanfeld, V. and Larson, E. L. (2010) Dissemination of health information through social networks: Twitter and antibiotics. *American Journal of Infection Control*, 38(3): 182–188.

Shirky, C. (2008) *Here Comes Everybody: The Power of Organising Without Organisations*. London: Allen Lane.

Simonsen, M., Skipper, L. and Skipper, N. (2015) Price sensitivity of demand for prescription drugs: Exploiting a regression kink design. *Journal of Applied Econometrics*. 31(2): 320–337.

Sornette, D., Deschâtres, F., Gilbert, T. and Ageon, Y. (2004) Endogenous versus exogenous shocks in complex networks: An empirical test using book sale rankings. *Physical Review Letters, American Physical Society*, 93: 228701.

Tesco PLC (2014) Major new partnership announced between Tesco, Diabetes UK and the British Heart Foundation. Available at: www.tescoplc.com/news/news-releases/2014/major-new-partnership-announced-between-tesco-diabetes-uk-and-the-british-heart-foundation/ (accessed 15 October 2016).

Twitter Inc. (2016) Getting started with Twitter. Twitter Support and FAQs. Available at: https://support.twitter.com/articles/215585 (accessed 31 July 2016].

Ulijaszek, S. J. and McLennan, A. K. (2016) Framing obesity in UK policy from the Blair years, 1997–2015: The persistence of individualistic approaches despite overwhelming evidence of societal and economic factors, and the need for collective responsibility. *Obesity Reviews*, 17(5): 397–411.

van Dijck, J. (2009) Users like you? Theorizing agency in user-generated content. *Media, Culture & Society*, 31(1): 41–58.

Wasilewska, E. (ed.) (2013) *Anthropology of Humour and Laughter*. San Diego: Cognella.

World Health Organization (2010) Diabetes fact sheet (February), p. 2. Available at: www.who.int/nmh/publications/fact_sheet_diabetes_en.pdf (accessed 15 October 2017).

Zimmer, M. and Proferes, N. J. (2014) A topology of Twitter research: Disciplines, methods, and ethics. *Aslib Journal of Information Management*, 66(3): 250–261.

4 Digital connections

Coffee, agency, and unequal platforms

Sarah Lyon

In 2015, the International Women's Coffee Alliance (IWCA) convened its fourth biennial international conference at the Corferias Convention Center in downtown Bogotá, Colombia. The multi-day event was timed to coincide with the ExpoEspecialies Café de Colombia, Colombia's annual specialty coffee convention. The IWCA conference featured coffee and international development industry speakers from around the world and was primarily conducted in English with simultaneous Spanish translation available via headset. In order to fill the cavernous event hall with bodies, the Colombian Coffee Growers Federation bussed in a large contingent of female coffee farmers from the Huila, Tolima, and Cauca regions. These women, very few of whom spoke English, were dressed in matching red vests and sat clustered together in the centre of the auditorium, chatting to one another and speaking on their cell phones throughout the day's presentations. On the afternoon of the second day, Ric Rhinehart, the Executive Director of the Specialty Coffee Association of America, presented detailed information about the evolving US coffee market. For the first time in two days the Colombian producers in the audience devoted their full attention to a speaker when Rhinehart began to discuss the possibilities technology holds for forging connections between producers and consumers. He noted that US consumers, especially those purchasing certified coffees (Fair Trade, organic, etc.), have long been interested in the lives of their producer correlates. Yet, he explained, 'Producers also have the right to know where their coffee is going, not just the other way around.' Rhinehart suggested that Skype is a tool producers could use to talk to coffee consumers in real time: 'The Internet will shrink the world to the point where consumers 10,000 miles away are truly as close as your neighbours.' The producers in the audience began clapping and cheering enthusiastically, clearly excited by the thought that they could translate digital connections with consumers into increased sales in a competitive marketplace.

Today virtually every real-world object has an 'information shadow'[1] in cyberspace that can be tracked by interested parties. The practice of traceability, facilitated by emerging digital tools such as radio frequency identification (RFID) tags which are used to track consumer products and digital archives of sales contracts,[2] is now common in the agri-food sector and ultimately it is about visibility – about revealing the production history of food to interested consumers. Increasingly the digital age and the ease of Internet access through mobile technologies are opening up possibilities for consumers to push beyond certification regimes and government regulation through tracking, rating, and researching the products they purchase using online platforms such as The Good Guide[3] (Lyon, 2014a). Less recognized, yet equally important, is the potential that digital technologies hold for producers, such as the Colombian coffee growers, who are interested in better understanding and influencing the marketplace.

The specialty coffee agri-food sector has long been at the forefront of traceability and consumer-producer connectivity, with its increasing attention to origin and the influence of certification regimes. Coffee's 'first wave', epitomized in the Colombian National Coffee Federation's long-running Juan Valdez advertising campaign, was characterized by increasing consistency in coffee quality and consumption throughout the twentieth century. Coffee's 'second wave' in the United States began in the 1960s and the 1970s with the founding of Peet's Coffees and eventually similar specialty coffee roasters, such as Starbucks, which introduced the American public to espresso products and single-origin coffees (rather than the blends that previously had predominated). The Specialty Coffee Association of America (SCAA) was founded in 1982 and slowly began to systematize the specialty coffee market through cupping standards and formal definitions. For example, specialty coffee in the green bean (pre-roasting) stage must score above 80 on the SCAA's 100-point scale and must be free of defects. Second-wave coffee is characterized by differentiated flavour profiles and often offers certified qualities such as Fair Trade, organic, or shade-grown (bird-friendly).

These trends are intensified within the emerging 'third-wave' coffee market, which consists of high-quality, single-origin, artisanal coffees that are being reimagined through an increasingly more sophisticated and complex set of distinctions such as terroir, ethical certifications, producer identities, micro-lots, quality classifications, roasting techniques, and brewing innovations. Coffee's recent 'third wave' presents opportunities to small-scale coffee farmers that can only be capitalized upon with a firm understanding of market demands and strong networks of buyers. While affluent urban US consumers do not think twice before paying $5 or more for a pour-over cup of specialty coffee, producer profit margins remain small and in many years returns barely cover the costs of production. Those coffee producers who are able to successfully brand their product or forge direct relationships with buyers are able to sell their coffee for consistently higher prices. Understandably, many producers are eager to develop a digital presence in order to strengthen the transnational connections that might give them a competitive edge in a crowded market.

This chapter explores three principal ways that digital technologies are employed by small-scale, Fair Trade/organic coffee farmers and their producer organizations in Southern Mexico, where Internet access is often more affordable and easily obtained (through Internet cafés and Wi-Fi on mobile devices) than phone service. It situates coffee farming practices within a broader understanding of producer/consumer politics within alternative food networks. It also explores how small producers are using digital tools to position their coffee within the growing third-wave market which privileges high-quality, single-origin, distinctive coffees. First, coffee producer organizations meticulously trace the coffee produced by members from field to cup in order to create micro-lots of premium coffee, or a small amount of coffee from a single farm or group of farms that is separated from general production and typically earns higher prices due to quality and flavour traits. These traceability practices also enable management to identify production and processing issues on farms before they begin to negatively impact the organization's coffee quality as a whole. Second, coffee farmers and producer organizations use the Internet to research coffee buyers and potential new markets, and service providers. Third, producer organizations employ social media to share their lived experience of coffee production and to initiate conversations with diverse constituencies, including consumers, buyers (roasters and importers), members, and funders. Some producer organizations also use social media to articulate political-economic viewpoints and lobby for policy changes.

Digital tools enable farmers to move beyond simple sound bites and photos mediated by coffee company marketing teams to share the practice of coffee production and their daily lives with distant others in ways that are more meaningful and resonant to the producers themselves. Scholars have long recognized the contemporary importance of consumer politics and how citizens creatively use the market as an arena for political action (Stolle and Micheletti, 2015). Some producers likewise engage in parallel forms of production politics, creating alternative markets and new types of economic transactions dependent on complex practices of traceability in order to influence government policies and consumer action. Production politics encompass the conscious agro-ecological choices farmers make (by farming organically or in ways that promote biodiversity), their participation in producer organizations, their active attempts to research and adapt to changing market demands, and the forms of political action and lobbying they engage in to pressure governments and organizing bodies to better meet their needs. As explored below, the Internet does indeed open up new possibilities for producers to assert their political agency and attempt to capture a higher percentage of coffee profits at origin. However, the producers' digital presence does not always conform to our own romantic images of coffee farming and the 'personal relationships' between producers and consumers supposedly facilitated by regimes such as Fair Trade (see Counihan and Siniscalchi, 2014: 10). By its very nature, the Internet articulates new geographies and is seemingly a perfect tool for cementing the transnational connections that make alternative markets such as Fair Trade possible. Yet, the coffee producers in Southern Mexico use the Internet to reinscribe their own place-based identities, articulating a regional vision for food sovereignty and social justice.

ICTs (information communication technologies) have been proven to revitalize communicative action in the public sphere and thus enhance participatory democracy (Carty, 2010). For example, Graham and Haarstad (2011) argue that the Internet creates alternative public spheres, or convergence spaces where both producers and consumers can practise new forms of transnational food citizenship. However, as this chapter highlights, producers and consumers have unequal platforms for action and there is a limit to the influence that the former can wield within markets, even when using these digital tools.

Research site and methods

This chapter features research data gathered among small coffee producers in the Southern Mexican state of Oaxaca, one of Mexico's most geographically and culturally diverse regions. Coffee, introduced to Southern Mexico by the Spaniards in the eighteenth century, is produced in mesophilic 'cloud-forest' regions, including the southern and northern mountains as well as the Mixteca. Coffee production in the region was well supported by the Mexican Coffee Institute (INMECAFE) until it disintegrated in 1989. At the same time the International Coffee Agreement, which long had stabilized international coffee prices, was dismantled. Consequently, coffee production in Mexico began to decline. In subsequent years, 71 per cent of Mexican coffee producers ceased to use chemical fertilizers and pesticides (Renard, 2010) due to their prohibitive costs. The nation's small producers, such as those in Oaxaca, were therefore well positioned to take advantage of expanding Fair Trade and organic coffee markets in recent decades. Mexico is the world's largest producer of organic certified coffee (USDA, 2010). Unfortunately, many coffee producers in the region are currently suffering from a recent outbreak of coffee rust. For example, the Coordinadora Estatal de Productores de Café del Estado de Oaxaca (CEPCO) estimates that there was a 40 per cent drop in coffee production state-wide during the 2014–2015 production cycle due to the spread of coffee rust (CEPCO, 2015). Coffee rust (*Hemileia vastatrix*) is a fungus that covers tree leaves with orange spores, eventually killing entire fields of coffee. While plantations can sometimes be resuscitated through aggressive pruning, producers frequently must replant with new, rust-resistant varietals.

Coffee is predominately a smallholder crop, produced in small communities that dot the mountainous terrain. Oaxaca is more than 50 per cent indigenous (with 16 different ethno-linguistic groups) and home to 15 per cent of Mexico's estimated 12 million indigenous citizens (Danielson and Eisenstadt, 2009). Although coffee cultivation is a household activity, it is mostly produced on collectively-held land. The average Oaxacan farmer has 1.24 hectares planted in coffee (CEPCO, 2015). Smallholders in Oaxaca typically wash, ferment, depulp and dry their coffee within their own homes and then sell the resulting coffee in parchment to buyers working for local mills and exporters or a coffee producer organization. Fair Trade and organic premiums are only available to those producers who belong to the latter. Many of Oaxaca's smallholder coffee producers have joined agricultural cooperatives in order to obtain Fair Trade and organic certifications and collectively sell their coffee to exporters and coffee roasters for higher prices than the coyotes, or local buyers, offer. These organizations were formed after the 1989 dismantling of INMECAFE and their long-term success results in part from the high levels of social capital found in many rural

Oaxacan communities, including collective control and sustainable management of natural resources; reciprocal and mutually supportive work systems; strong social organization and high levels of communal responsibility; and a deep respect for the knowledge of community elders (Hall and Patrinos, 2012: 7).

The research was conducted between 2014 and 2016 and included a survey of 489 coffee producers across four different regions in the state of Oaxaca and focus groups with the members of each of the five participating coffee producer organizations. In addition, the research team conducted multiple informal, unstructured interviews with cooperative members, management, and staff, and engaged in participant observation at cooperative meetings and in the communities. Prior to beginning research in each community we sought formal approval from the mayor's office and also secured written permission from the management of each participating coffee organization. While we obtained a waiver of documentation of informed consent from the institutional review board, verbal informed consent was secured from each participating individual; each participant also received a written document including information about the research goals and contact information for future questions and concerns.[4]

Digital connections in Mexico and beyond

Online platforms are built on the assumption of open, communicative access, yet most often in practice they cater to a privileged demographic whose agenda and viewpoints prevail (Dean, 2009). Although historically true, as Internet technologies spread across the globe, these power relations may begin to shift in favour of the disadvantaged: while wider usage does not necessarily ameliorate power imbalances, it does help people, even in rural, 'out-of-the-way' places to incorporate digital tools into their own daily lives and communities of practice. Sixty-three per cent of Mexicans own cell phones and 27 per cent of these users access the Internet via their phones (Pew Research Center, 2013). On the other hand, only 35 per cent of Mexican households owned a computer in 2010 and the penetration of computers in Mexican households at the top of the economic pyramid was 5.5 times that of households at the bottom (Villagran, 2012). Smartphone usage in Mexico has doubled annually between 2012 and 2014 and there are now an estimated 33.3 million smartphone users, representing one-quarter of the country's population (Oleaga, 2014).

There is often a tendency to portray cyberspace as a placeless, culture-less arena (Miller and Slater, 2000), yet these technologies are actually integrated into the lives of real people in very specific ways. The inhabitants of Oaxaca's rural farming communities use a surprisingly high number of Mexico's smartphones, as typically an entire village will share one communal landline (with people's names announced over fuzzy megaphones when a call awaits them). Farmers frequently use cell phones to access the Internet via wireless connections offered by Internet cafés or local businesses even when they cannot actually receive calls or text messages due to the region's mountainous terrain. Therefore, while few (if any) individual producers actually own computers, many do regularly access the Internet and social media, such as Facebook, through phone applications. Fair Trade coffee producer organizations typically have at least one computer and regular Internet access. If the latter is beyond the organization's means, then at the very least a business manager regularly visits Internet cafés for email correspondence.

The politics of producer-consumer connectivity in Fair Trade coffee markets

Histories of the Fair Trade movement often date its origins to the post-World War II era when alternative trade organizations started importing artisan products into Europe and North America. However, the roots of the contemporary certification-based Fair Trade market for agricultural commodities in fact lie in Oaxaca's coffee-producing communities. In 1983, farmers from 17 indigenous Oaxacan communities formed the Union of Indigenous Communities of the Region of the Isthmus (UCIRI) in an attempt to address the fact that they were not receiving fair prices for their coffee. In 1988, they submitted a proposal to Solidaridad, a Dutch development organization, which resulted in the formation of Max Havelaar, the world's first Fair Trade certifying body (VanderHoff Boersma, 2009). Simultaneously, in 1985, the Catholic Church's local cooperative commission held a seminar with Maya coffee growers in Chiapas, Mexico, to reflect on problems and seek possible solutions, resulting in the formation of the Indígenas de la Sierra Madre de Motozintla (ISMAM) cooperative. In working with their partners in consuming countries, the Mexican coffee cooperatives offered an alternative vision for creating a just society 'in which markets provide a means for meeting human needs rather than defining the essence of what it is to be human' (ibid.). This ethic clearly influenced the ultimate goals and practices of the emergent certification-based Fair Trade movement which has increased dramatically in scale and scope over the past two and a half decades.

As Goodman *et al.* (2012: 209) point out, Fair Trade has been at the forefront of developing the possibilities of Internet technologies as tools for building interactive connections between consumers and Fair Trade producer communities and, in recent years, the transparency Fair Trade networks has moved into virtual spaces in ways that far eclipse the simple inclusion of descriptive text and photos on coffee bags. For example, Fair Trade consumers can now use the Internet to research coffee producer organizations in-depth, learning their unique histories and details about local culture. It is usually possible to identify how long an organization has been Fair Trade certified, other certifications they may have, and the importers who work as intermediaries between producer organizations and coffee roasters. Frequently, specific information about contract negotiations and pricing information is also readily obtainable. This differs from earlier forms of Fair Trade visibility, which were typically limited to a stock photo and a short description written by a member of the coffee firm's marketing department. However, even today, this digital transparency remains largely one-sided and highly structured by the power inequities that shape the coffee market itself:

consumers have the privilege of 'seeing' and even knowing coffee producers but the latter have less opportunity to see who is buying their products, how, and why (ibid.: 220). Given that consumer politics within alternative agri-food networks (formed by groups of consumers, producers, and advocates who work to cultivate food relations that move beyond simple market transactions) have been so influential in recent years, it is not surprising that scholarly analysis has centred on consumer power. Yet, it is equally important to explore the efforts of the producers who have inspired and often created many of the alternative agri-food initiatives, such as Fair Trade , in existence today (Allen, 2014: 75).

Third-wave coffee, traceability, and the politics of quality

As noted earlier, third-wave coffee, emerging over the past decade, is characterized by a growing focus on quality and qualities. Third-wave coffee privileges the individualization of the coffee experience, for example, through pour-over, single-cup servings rather than cups filled from pre-brewed pots, and the increasing sophistication of coffee shop culture and barista skill. The latter is regularly celebrated through competitions such as the U.S. Barista Championship. Third-wave coffee is also noteworthy for an obsession with coffee provenance, not simply country of origin but often a detailed accounting of specific farms and their terroir (the environmental characteristics that influence coffee's epigenetic qualities). Distinguished coffees are often sold at high prices in micro-lots and demand for specific coffees is heightened through regional Cups of Excellence competitions. It is important to note that coffee's successive waves are defined by shifts in marketing and consumer sensitivities, rather than production-related issues, highlighting the power imbalances that continue to structure coffee commodity networks.

Krzywoszynska (2015) maintains that the marketization of alternative or quality foods, such as third-wave coffee, involves not only strategic positioning, trust, and cultural mediation, but also the cultivation of consumers' taste as an experientially shaped form of visceral attachment. As Vercelloni points out (2016), 'taste' emerged as a hallmark of consumer society when capitalism was transformed, beginning in the nineteenth century, from a mode of production rooted in self-discipline and sensory restriction into a mode of seductive consumption premised on self-indulgence and sensory enticement (Howes, 2016: xii). Third-wave coffee and its focus on quality, artisanal characteristics, and distinctive flavour profiles depend entirely on the further refinement of consumer tastes. It is also characteristic of the larger 'quality turn' in food systems (quality of work, health, and life), which emerges from a sense of discontent with the industrial food system and a turn instead towards the ' "interpersonal food world" where quality conventions embed trust and tradition within a moral economy of place and provenance' (Constance et al., 2014: 22). Traceability is a key component of the broader quality turn as it enables consumers to make informed food choices and serves as a vehicle for those who wish to actively participate in the reshaping of our global and local food supply. At the same time, producers can also use traceability to ensure the veracity of the qualities they market to consumers.

For example, Café de Oro,[5] a large Oaxacan coffee cooperative located in the south-western part of the state, uses a detailed traceability protocol in order to separate micro-lots of high quality coffee produced by some of its women members for sale to a roaster in the United States, which in turn markets the gendered quality of the coffee alongside its organic and Fair Trade certifications. Traceability is critical because, alongside these micro-lots, the cooperative also separates its organic certified coffee for sale to US roasters (through an importer), its transition (organic coffee that is not yet certified) coffee for sale to an Oaxaca City-based coffee roaster, and its conventional (non-certified) coffee for in-house roasting and local markets. The micro-lots of women's coffee begin their journey from field to cup in women's homes where coffee cherries are processed, fermented, and dried. The resulting parchment coffee is collected in burlap sacks and carried to the local committee's office where it is weighed and tagged before storage. A 3-inch green index card is stapled to every sack of parchment coffee turned in by members, listing its community provenance and the producer's name (both of which can be cross-referenced for altitude and certification status in the cooperative's computer records), the date (and whether this was the first, second, or third batch of coffee turned in), and the number of total sacks turned in by the producer on that day.

Café de Oro has 711 members living in 25 different communities, all at very different altitudes and with microclimates that shape the coffee's epigenetic qualities. Consequently, once the parchment coffee reaches the cooperative's large warehouse, it is strictly organized according to community and gender. All of the coffee produced by the men in one community or the women in another will be processed in batches and this community/gender identity will remain with the coffee once it is processed. Each sack of green coffee is labelled and cross-referenced in digital records before being loaded onto a truck for a journey over the mountains to the port in Veracruz from where it will head, by boat, to New Orleans. The carefully labelled coffee will then be trucked to individual buyers throughout the United States, each of whom will use digital tools, such as RFID tags, to continuously trace the coffee through the shipment, storage, roasting, and sale stages. This entire process enables third-wave coffee roasters to purchase micro-lots of coffee that have exceptional flavour profiles or are produced by women and to make visible to consumers this production history. It is also illustrative of the digital system's tendency to intensify the dialectic between universality (every bag of coffee can be identified through the same process, with identical tags) and particularity (each bag is produced by an individual farmer in a specific location) (Horst and Miller, 2012).

The traceability helps Café de Oro to sell their coffee for slightly higher prices and, in essence, the traceability becomes a form of rent within the commodity network. It also enables the cooperative to identify and isolate potential production and processing issues. For example, if the majority of the coffee produced in one community presents with a sour or winey smell (produced by over-fermentation), management may encourage the residents to take out small loans to buy new depulpers to decrease processing time.

This particular coffee commodity chain is organized and managed through an assemblage of digital and non-digital tools, such as the green paper tags affixed to bags of green coffee when they are turned over to the producer organization by an individual farmer; the computer records of the organization which uses these tags to track the coffee through the processing stages and comply with certification requirements; email communication between the producer organization, its exporter, and coffee roasters in the US; and the RFID tags affixed to bags of coffee prior to retail sale. While ultimately the coffee's traceability forms part of the vast 'Internet of Things', or the coding and networking of everyday objects and things to render them individually machine-readable and traceable on the Internet (Graham and Haarstad, 2011), its base rests on the seeming old-fashioned tools of pencils, paper, and book keeping. Graham and Haarstad (ibid. 2) argue that the Internet of Things can potentially create a different type of globalization, 'one characterized by knowledge and transparency and able to harness the power of the Internet to allow consumers to learn more about the commodities that they buy.' However, as they point out, this potential is forestalled by barriers to the creation and transmission of information about commodities, including infrastructure and access, actors' capacities, and the continued role of intermediaries. For the potential to be realized, technological change must be embedded in broader processes of local capacitation, democratization, and social change (ibid.: 2), a theme that will be returned to in the conclusion.

The US roaster, which purchases the majority of Café de Oro's women's coffee, sells it for a premium on the shelves of a nation-wide specialty grocery store. The roaster pays the cooperative a small premium for this high-quality coffee and the money is used to pay the salary of a women's programme coordinator and to assist women members through programmes such as revolving credit funds and kitchen garden projects. On store shelves there is no signage explaining the women's project and very little text on the actual coffee bags so that, other than the blend's name, which identifies it as women-produced, very little information is available to consumers. Instead, the Internet is the key platform for marketing this women's coffee and the roaster advertises on its website how:

> we noticed over the years that women's coffee often outshines the men's. That, and the fact that historically female coffee growers received little recognition for their hard work, we decided to ask that beans be separated to make a women's super-lot.

> (Allegro, 2015)

In a 2013 blog post (FTUSA, 2013), the roaster's Director of Sourcing and Quality Control explained how Café de Oro originally proposed the idea of the Women's Coffee, describing it:

as a way to garner a higher premium and provide an opportunity for a higher level of participation in the co-op by the female members, many of whom were taking care of their children alone while their husbands were working in the United States. After having their lots separated and bulked into full container volumes for a couple of years, we noticed that these women's lots consistently scored a couple points higher than the other co-op lots that we bought. This outcome tied back nicely to the idea that women are generally better care givers, and this carries over to the added attention they pay to the horticulture, harvesting and processing of coffee.

This attention to quality in both senses of the word (e.g. high-quality coffee that is embedded with gendered qualities such as care-giving) is a hallmark of third-wave coffee.

Cooperative leaders and managers obviously appreciate the premium the roaster pays for the coffee, however, many members were either not aware of the programme or were under the mistaken impression that the price premium would be paid back to individual producers. Significantly, only 56 per cent of women and 86 per cent of men were actually aware of Café de Oro's micro-batching programme, yet there was widespread support for these types of micro-batching initiatives in theory: 96 per cent of surveyed members said they were in favour of special sales for women's coffee. When I asked the women's program-ming coordinator how the women's coffee programme could be enhanced, for example, through a higher premium, she responded: 'The members have a real desire to meet directly with the roasters and to learn more about them and coffee commercialization.' While producers know that they must produce high-quality coffee in order to earn high prices, they are not always clear what this entails beyond simply following organic certification requirements. The women are interested in learning more about how their coffee is marketed and how they can build on its success. Through traceability and direct sales, both of which are facilitated by the Internet, the cooperative clearly understands who is buying their coffee and how they are marketing it to consumers. However, the members' desire to meet with the roasters highlights the fact that digital information on its own is not sufficient: it must also be easily assessed as reliable and accurate, salient, and relevant for those making decisions or choices, and credible (Hepting *et al.*, 2013). In other words, it is not enough to have access to this information: producers must be able to trust it and situate it within a meaningful contextual framework. Whether a meeting between Café de Oro members and the coffee roaster might be facilitated through technology (e.g. Skype), as Ric Rhinehart suggested to the Colombian coffee growers, remains to be seen.

Researching and reaching out to coffee buyers via the Internet

Another Oaxacan cooperative we worked with was recently formed when a group of younger coffee farmers left a well-established coffee organization out of frustration with its perceived corruption and ineffectiveness. These producers, who primarily hailed from a single community, formed their own group and subsequently joined the state's largest second-tier coffee producer organization. The leaders are quite savvy and strategic – they joined the second-tier organization because of its strong and secure markets and the wide array of non-coffee-related services it offers. For example, the second-tier organization helps members of its affiliated groups to amass the paperwork and documentation needed to participate in the Mexican government-funded housing improvement programme. The members also recognize that due to their community's high altitude, their coffee has a unique and desirable flavour profile. However, Maria (a pseudonym), who served as a leader in the organization she and others eventually left, explained to me:

> One of the main challenges I see is that people need to learn more about coffee commercialization – where does it go after they grow and process it? Where is it dry milled? Where is it roasted? When I was serving as a representative, they sent a delegation to the roasters in the US to learn more about this and I really wanted to go on this trip but unfortunately I wasn't chosen.

One of the organization's long-term goals is to export the coffee grown in their community, Cabeza del Rio, under that name. Maria sees that some other communities are able to do this and wonders why they cannot as well. 'Why not?' she asks, 'If others can do it, then we should be able to do it too.' Since opportunities to travel to the United States and meet directly with buyers are extremely limited (and, even when they do present, the competition within organizations for the available slots is often steep), Maria used the Internet to research Oaxaca-based coffee buyers who export to the United States and reached out to them with offers to send samples of green coffee. The Internet also revealed to her the ways in which other communities market their own coffees and how events such as the Premio a la Calidad del Café de Oaxaca celebrate the region's unique flavour profiles. This event, organized over the Internet and convened each spring by the Autonomous University of Chapingo, is modelled on the Cup of Excellence competitions and winners (listed by name, community, and municipality) from each of Oaxaca's coffee-producing regions are chosen through cupping. The winner of the 2016 competition, Vicencio Gracida Hernandez, sold his green coffee for $16/pound (Premio, 2016), in April 2016 when the international coffee market price was only $1.30/pound. This example highlights how the Internet enables producers to see what has previously been invisible to them.

While Maria and her colleagues certainly use the Internet for market research, they have yet to actively advertise their own coffee or organization to consumers. However, this is a step that an increasing number of coffee organizations, even smaller ones, are now taking. For example, another organization in the region, Yuku Kafe, has a frequently updated Facebook page (Yuku Kafe, 2016) that highlights not only coffee-related activities but also community events and workshops for local residents. The Yuku Kafe project, funded by a grant from the Vibrant Village Foundation (a Portland, Oregon, community development organization), promotes coffee grown in the Mixteca Alta region to international buyers and aims to increase coffee quality and yields through training and workshops. The aim is to identify high-quality producers and support them in their efforts to connect with buyers. Recognizing the critical role that viable coffee production plays in sustaining rural Oaxacan communities, the organization also aims to educate future generations of coffee farmers and present them with an alternative to migration. In addition to social media (Yuku Kafe also has a Twitter account), the organization has created demonstration plots and encourages visitors to come and witness high-quality coffee production in process. It also hosts events such as a June 2016 public cupping of micro-lots in Mexico City.

Web 2.0 platforms, such as Facebook, centre on user-generated content and offer a number of advantages for fostering collective action in comparison to offline alternatives. For example, it is much easier to discover and attract members (or followers – Yuku Kafe has over 700 likes) with shared interests; to exchange information; to integrate individual contributions; and to manage group logistics (Linders, 2012). What makes the Facebook pages of coffee producer organizations such as Yuku Kafe so fascinating is the multi-faceted and two-way nature of the communication they advance. Yuku Kafe's page features links to articles and blog posts written in English and Spanish that discuss aspects of production and processing that impact coffee quality, notices of coffee-related conferences, advertisements for the organization's own events (such as the Mexico City cupping), photos of organizational activities, and information about upcoming workshops. The digital platform allows the group to speak to multiple constituencies simultaneously – coffee buyers, consumers, and members – and it allows these individuals to 'speak back' through comments and replies. It is representative of the types of 'convergence spaces' fostered by the Internet and characterized by a decentralized and non-hierarchical structure, immediate solidarity, communication, and alliance-building across space, and a diffuse networked force that challenges neoliberal globalization (Graham and Haarstad, 2011: 5).

Sharing daily life and claiming justice via social media

A critical component of crafting a successful digital presence is maintaining a consistency of message across diverse spaces and communities. The second-tier coffee organization CEPCO has one of the most active coffee-related social media presences in Mexico. The organization's Facebook page is updated almost daily and, much like Yuku Kafe, it includes photos from activities and meetings organized by CEPCO itself and by the 36 coffee producer organizations that belong to CEPCO. It regularly features reports and photos of the lived experience of coffee production and specifically highlights coffee renovation efforts related to coffee rust mitigation, for example, photos from greenhouses showing coffee seedlings or descriptions and photos of workshops on the proper use of copper sulphate to treat coffee rust. Like Yuku Kafe, CEPCO uses its Facebook page to communicate with coffee consumers, buyers, donors, and, importantly, members. Maintaining regular contact and sharing information with coffee producers who are spread out across Oaxaca's rugged terrain is a perennial challenge for coffee organizations. Local cooperative representatives typically spend an entire day travelling to Oaxaca City for CEPCO's monthly meetings, sleep on cots in the CEPCO offices, and then spend a day returning to their communities. After spending three or four days travelling, they are eager to return to their own coffee fields and work activities and often do not have the time to subsequently travel to distant communities (which might be 8–10 hours away) to share information and update members. At one monthly meeting of the organization's board of directors (comprised of delegates from the 36 member organizations) in Oaxaca City, CEPCO's director urged the representatives in attendance to set up their own organizational Facebook accounts to communicate important information to their members. At least a handful of the member organizations have done so, however, their pages are not nearly as active as CEPCO's itself.

What is unique about CEPCO's social media account is that amidst these portrayals of the daily practices and business of coffee production, the organization articulates a clear political vision and regular calls for social justice. These take the form of links to petitions: a recent post linked to one urging Starbucks to dismantle its Todos Sembramos Café programme, which pledged to donate a coffee plant to farmers in Chiapas, Mexico, for every bag purchased (see Starbucks, 2014).[6] It also features more mundane forms of political expression, such as a drawing of Zapatistas drinking coffee captioned 'Nadie se Rinde' (no one surrenders). The Facebook page also serves as a platform for formal political statements such as a 20 June 2016 post featuring a formal condemnation, signed by CEPCO's member organizations, of the federal government's violence against members of the independent teacher's union, the National Education Workers Coordinating Community. The majority of the organization's political messaging centres on concerns relevant to campesinos, or small producers, in Mexico and beyond. For example, reminding people that small producers grow 40 per cent of the food eaten in the world, that small coffee farmers play a critical role in protecting biodiversity, and that the future of coffee production in the region is in doubt due to climate change.

CEPCO leaders maintain a great deal of consistency in the messages they share online and during in-person encounters, both at monthly meetings and at events in producer communities. For example, CEPCO has been a vocal critic, along with many other farmer organizations within Mexico, of the federal government's lack of support for small farmers and, in particular, its failure to offer assistance to those facing significant losses due to the coffee rust outbreak. In May 2015, one of CEPCO's member organizations in the Loxicha region celebrated its 25th anniversary, a festive event that several of the organization's Oaxaca City-based staff attended. In his speech, one member of the managerial team acknowledged the local cooperative's successful history, saying, 'We are happy that you have remained united. This is not easy.' He asked everyone in the audience to stand for a long moment of silence for those who had passed, especially those who had died 'during the revolution'. In 1996, the military and the police arrested all of the town's municipal authorities and initiated a lengthy campaign 'of what residents and human rights agencies describe as sheer terror' (Rochlin, 2007: 122), accusing them of supporting the Popular Revolutionary Army (or the EPR, a revolutionary group active in the region and unrelated to the Zapatistas (EZLN)). He went on to speak for nearly 45 minutes, detailing the many programmes and forms of support that CEPCO offers its member organizations and affiliated coffee producers, implicitly, and at times explicitly, contrasting these with the lack of support Mexico's federal government provides to contemporary campesinos: 'Well, if the government helps us, that would be great,' he said. 'But even if it doesn't, we will continue working to move forward.' This message is consistent with many of the posts on the organization's Facebook page that explicitly or implicitly criticize the government's discriminatory policies. For example, in July 2016, the organization posted a graphic produced by the Valor al Campesino campaign which criticizes SAGARPA's (The Secretariat of Agriculture, Livestock, Rural Development, Fisheries and Food) treatment of small producers (those with less than 5 hectares of land) in the southern part of Mexico (Valor, 2016) and it regularly uses Facebook to promote the activities of the Colectivo Huaxyacac, a collective of civil society and producer organizations working to lobby the state government to adopt more progressive environmental, social, and political policies.

There has been much collective hand-wringing over the future of Fair Trade in recent years, especially after Fair Trade USA decided to leave Fair Trade International in 2011 in order to adopt a more 'business-friendly' outlook and open up Fair Trade certification to plantation-produced coffee (since Fair Trade International limits Fair Trade certification to small farmer producers) (Lyon, 2014b). A primary tension within the Fair Trade movement has always existed between those who view Fair Trade as primarily providing a shaped advantage by assisting poor producers to participate in global markets vs. those who understand it to be an alternative to the neoliberal market model (see Fridell, 2007). However, it is questionable whether these debates accurately reflect the viewpoint of the participating farmers whose interest the Fair Trade movement ostensibly serves. As Allen (2014: 74) points out, the survival of producers within alternative food networks such as Fair Trade depends on market niches for products of a specific quality and

> they try to survive in the market without attempting to radically transform it, a goal they do not intend any more: perhaps certain criticisms of these niche strategies are the result of attributing to them intentions that were never in fact theirs.

Fair Trade promotes a view of farmers as the 'deserving poor' (Adams and Raisborough, 2008), who should be supported with fair prices for the goods they work so hard to produce. Polynczuk-Alenius and Pantti (2016) argue that portraying these vulnerable farmers as active, entrepreneurial agents (rather than as passive victims) offers Northern audiences greater possibility for sympathy and identification, however, Fair Trade's representation of producers as active subjects is somewhat ambiguous as their agency still depends on the ethical conduct of consumers who purchase the producers' coffee, thereby ensuring the continued viability of their livelihoods. They point out that social media platforms such as Facebook, with its 1.5 billion worldwide users, have the potential to accommodate a multiplicity of voices, raise awareness, and foster the direct and vocal participation of previously marginalized actors, who can now reach others by producing social media content (ibid.: 3). Yet, the examples of Yuku Kafe and CEPCO's Facebook pages highlight the different modalities of producer communication via social media: using the Internet to build a brand and position producers within a marketplace supports a very different form of agency than CEPCO's more explicit political messaging and calls for social justice within a specific regional context.

Farmer organizations such as CEPCO are indeed using social media platforms to participate in broader conversations, however, their articulated viewpoints focus less on the politics of Fair Trade and the inequities of global markets (topics that ethical consumers find compelling) and instead on topics more closely aligned with the daily life of coffee production and local political concerns. Their message is more closely aligned with discourses of food sovereignty which promulgates a 'relational, historically and culturally grounded "people-centred" approach [that] can highlight the social elements that create and/or strengthen resonant, locally inflected political strategies for food sovereignty' (Figueroa, 2015: 500). While a somewhat intentionally ambiguous, 'big tent' concept, food sovereignty is most commonly defined as

> the right of peoples to healthy and culturally appropriate food produced through ecologically sound and sustainable methods, and their right to define their own food and agriculture systems. It puts those who produce, distribute and consume food at the heart of food systems and policies rather than the demands of markets and corporations.
>
> (Via Campesina, 2007)

As Patel (2009) points out, when we talk about food sovereignty we essentially talk about rights and that means we must also discuss ways to ensure that those rights are met, across a range of geographies, in substantive and meaningful ways. In its online presence, CEPCO is doing precisely this: discussing the rights that Mexican campesinos have to engage in meaningful forms of agricultural work that are supported by their government and equitable terms of trade. For small-scale coffee producers questions about justice do not necessarily centre Fair Trade standards and requirements (e.g. debates over whether certification should be extended to plantations), but instead explore topics with more immediate relevance such as violence against workers' unions and lack of federal support for farmers facing substantial losses from coffee rust. As

Figueroa (2015: 505) points out, food sovereignty's radical imperative is to consider social justice as the very foundation of a sustainable food system: 'The primacy of the social in the ideas and goals of food sovereignty, therefore, seems to suggest a reorientation in the critique of food systems towards the social, and the organization of society as a whole.'

Conclusion

While the Internet holds a great deal of unrealized potential for small producers who wish to connect to distant consumers on their own terms, a possibility suggested in this chapter's opening vignette, an increasing number of marginalized farmers are incorporating it into their daily business and organizational practices with great success. In Oaxaca, Mexico, small coffee producers and their associations use digital tools such as online record keeping and email to meticulously trace the coffee produced by members from field to cup in order to create micro-lots of premium coffee and to identify production and processing issues on farms before they begin to negatively impact the organization's coffee quality as a whole. Coffee farmers and producer organizations also use the Internet to research coffee buyers and potential new markets and service providers. Finally, and perhaps most interestingly, producer organizations employ social media to share their lived experience of coffee production and to initiate conversations with diverse constituencies, including consumers, buyers (roasters and importers), members, and funders. More politically active groups carefully align their digital presence with specific visions of locally meaningful forms of social justice and calls for action. Each of these forms of digital labour (traceability, market research/branding, and political messaging) facilitates different degrees of producer agency. A key difference is that the first two are dependent on building connections with sympathetic buyers and consumers and hence are shaped by the larger power inequities structuring the international coffee market. The third form of digital labour, the social justice activism evident on CEPCO's social media account, helps create and nurture a particular political subjectivity at the organizational level. The organization exerts a great deal of agency in presenting its own self-image. Yet, given that the majority of posts are in Spanish and speak to regional, rather than international issues, this agency is limited within the larger coffee market itself.

Despite these inequities, the digital practices explored in this chapter do constitute 'convergence spaces' rather than formal networks and structured conversations. Graham and Haarstad (2011: 5) maintain that these convergence spaces are what are unique about Internet-enabled politics. Within the realm of food politics, convergence spaces become sites for the promotion of ideas of food citizenship, which Wilkins (2005) defines as the practice of engaging in food-related behaviours that support, rather than threaten, the development of a democratic, socially and economically just, and environmentally sustainable food system. However, as noted earlier, in order for technology to foster social change, it must be embedded in broader processes of local capacitation and democratization. Each of the examples explored demonstrates a degree of local capacity building and, given the increasingly widespread access to the Internet in rural Mexico, the tools it offers are certainly being democratized. However, producers must rely on existing platforms and do not have the capacity or resources to build their own, alternative platforms. Consequently, there is still a limit to the degree of agency producers can exert within the coffee market and given the existing inequities, it is unlikely that digital tools will enable them to permanently and thoroughly shift these power imbalances in the near future.

Notes

1 A product's digital information shadow includes any information readily accessed on the Internet, including place and conditions of production, pricing information (and, in coffee markets, sometimes sales contracts), certifications, and retail outlets. This information shadow parallels the digital footprint left by Internet users as a result of web-browsing, stored cookies, and social media activity.
2 For example, see Just Coffee's (Madison, Wisconsin) transparency project, which includes an archive of all sales contracts, available at: www.fairtradeproof.org/just-coffee/index.php?rst_id=10 (accessed 15 August 2016).
3 See www.goodguide.com/ for more information (accessed 15 October 2017).
4 The research was conducted by a team of investigators, including the author, Tad Mutersbaugh, and Holly Worthen, who had human subjects approval from the University of Kentucky's Institutional Review Board (IRB #14–0898-P4K).
5 A pseudonym.
6 The petitioners claimed that the programme lacks transparency. See https://secure.avaaz.org/es/petition/Howard_Schultz_dueno_de_Starbucks_Que_detenga_el_programa_Todos_sembramos_cafe_en_Mexico_y_Latinoamerica/?cHMmOkb (accessed 15 August 2016).

References

Adams, M. and Raisborough, J. (2008) What can sociology say about Fair-trade? Class, reflexivity, and ethical consumption. *Sociology*, 42(6): 1165–1182.

Allegro (2015) Organic Cafe La Dueña. Available at: www.allegrocoffee.com/coffee/coffee-products/coffee-blends/product/organic- cafe-la-duena (accessed 13 July 2015).

Allen, P. A. (2014) Divergence and convergence in alternative agrifood movements: Seeking a path forward. In D. H. Constance, M-C. Renard and M. G. Rivera-Ferre (eds) *Alternative Agrifood Movements: Patterns of Convergence and Divergence.* Bingley, UK: Emerald Group, pp. 49–68.

Carty, V. (2010) New information communication technologies and grassroots mobilization. *Information, Communication and Society*, 13(2): 155–173.

CEPCO (2015) Coordinadora Estatal de Productores de Café del Estado de Oaxaca A.C. Congreso XIII Informe de Actividades 2013–2015. Unpublished report. Oaxaca, Mexico: CEPCO.

Constance, D. H., Friedland, W. H., Renard, M-C. and Rivera-Ferre, M. G. (2014) The discourse on alternative agrifood movements. In D. H. Constance, M-C. Renard and M. G. Rivera-Ferre (eds) *Alternative Agrifood Movements: Patterns of Convergence and Divergence*. Bingley, UK: Emerald Group, pp. 3–48.

Counihan, C. and Siniscalchi, V. (2014) *Food Activism: Agency, Democracy and Economy*. New York: Bloomsbury Academic Press.

Danielson, M. S. and Eisenstadt, T. A. (2009) Walking together, but in which direction? Gender discrimination and multicultural practices in Oaxaca, Mexico. *Politics and Gender* 5(02): 153–184.

Dean, J. (2009) *Democracy and Other Neoliberal Fantasies: Communicative Capitalism and Left Politics*. Durham, NC: Duke University Press.

Figueroa, M. (2015) Food sovereignty in everyday life: Toward a people-centered approach to food systems. *Globalizations*, 12: 489–512.

Fridell, G. (2007) *Fair-trade Coffee: The Prospects and Pitfalls of Market-Driven Social Justice*. Toronto: University of Toronto Press.

FTUSA (2013) Fair-trade USA: Celebrating women's coffee. Available at www.fair tradeusa.org/blog/celebrating-women-coffee-mother-s-day-cafe-la-due-miracle (accessed 6 December 2015).

Goodman, D. E., DuPuis, M. and Goodman, M. K. (2012) *Alternative Food Networks: Knowledge, Practice, and Politics*. New York: Routledge.

Graham, M. and Haarstad, H. (2011) Transparency and development: Ethical consumption through Web 2.0 and the Internet of Things. *Information Technologies and International Development*, 7(1): 1–18.

Hall, G. H., and Patrinos, H. (2012) *Indigenous Peoples, Poverty, and Development*. New York: Cambridge University Press.

Hepting, D. H., Jaffe, J-A. and Maciag, T. (2013) Operationalizing ethics in food choice decisions. *Journal of Agricultural Environmental Ethics*, 27(3): 453–469.

Horst, H. A. and Miller, D. (2012) *Digital Anthropology*. London: Berg.

Howes, D. (2016) Preface: Accounting for taste. In L. Vercelloni, *The Invention of Taste: A Cultural Account of Desire, Delight, and Disgust in Fashion, Food, and Art*. Trans. K. Singleton. New York: Bloomsbury Academic Press.

Krzywoszynska, A. (2015) Wine is not Coca-Cola: Marketization and taste in alternative food networks. *Agriculture and Human Values*, 32: 491–503.

Linders, D. (2012) From e-government to we-government: Defining a typology for citizen coproduction in the age of social media. *Government Informational Quarterly*, 29: 446–454.

Lyon, S. (2014a) The Good Guide to 'good coffee'. *Gastronomica*, 14(4): 60–69.

Lyon, S. (2014b) Fair-trade towns USA: Growing the market within a diverse economy. *Journal of Political Ecology*, 21: 145–160.

Miller, D. and Slater, D. (2000) *The Internet*. Oxford: Berg.

Oleaga, M. (2014) Mexico tops smartphone market in Latin America with 50 per cent growth in 2013, becomes interest for mobile ad makers. *Latin Post*, 6 February 2014. Available at: www.latinpost.com/articles/6946/20140206/mexico-tops-smartphone-mobile-market-latin-america-50-percent-growth.htm (accessed 1 June 2016).

Patel, R. (2009) What does food sovereignty look like? *Journal of Peasant Studies*, 36(3): 663–706.

Pew Research Center (2013) Cell phones nearly universal in much of the world. Washington, DC: Pew Research Center. Available at: www.pewresearch.org/daily-number/cell-phones-nearly-universal-in-much-of-the-world/ (accessed 27 May 2016).

Polynczuk-Alenius, K. and Pantti, M. (2016) Branded solidarity in Fair-trade communication on Facebook. *Globalizations*. Published online 28 April 2016.

Premio (2016) Premio a la Calidad de Cafe de Oaxaca, 3rd edn. Available at: www.facebook.com/384564538346852/photos/pcb.764871310316171/764871180316184/?type=3& theater (accessed 3 June 2016).

Renard, M. (2010) The Mexican coffee crisis. *Latin American Perspectives*, 37(2): 21–33.

Rochlin, J. (2007) *Social Forces and the Revolution in Military Affairs*. New York: Palgrave Macmillan.

Starbucks (2014) Helping coffee farmers thrive in Chiapas, Mexico. Starbucks Newsroom, 17 June 2014. Available at https://news.starbucks.com/news/how-starbucks-is-helping-chiapas-coffee-farmers-thrive (accessed 17 June 2016).

Stolle, D. and Micheletti, M. (2015) *Political Consumerism: Global Responsibility in Action*. Cambridge: Cambridge University Press.

USDA (2010) GAIN Report, Mexico, 7 May 2010, XM003. Available at: http://gain.fas.usda.gov/Recent%20GAIN%20Publications/Coffee%20Annual_Mexico%20City_Mexico_5-24-2010.pdf. (accessed 17 October 2015).

Valor al Campesino (2016) Available at: https://valoralcampesino.org/ (accessed August 16, 2016).

VanderHoff Boersma, F. (2009) The urgency and necessity of a different type of market: The perspective of producers organized within the fair-trade market. *Journal of Business Ethics*, 86(S1): 51–61.

Vercelloni, L. (2016) *The Invention of Taste: A Cultural Account of Desire, Delight, and Disgust in Fashion, Food, and Art*. Trans. K. Singleton. New York: Bloomsbury Academic Press.

Via Campesina (2007) Nyéléni Declaration. Sélingué, Mali. Forum for Food Sovereignty. Available at: http://viacampesina.org/en/index.php/main-issues-mainmenu-27/food-sovereignty-and-trade-mainmenu-38/262-declaration-of-nyi (accessed 19 June 2016).

Villagran, L. (2012) For most Mexicans, the digital age is still out of reach. *Dallas Morning News*. 16 January. Available at: www.dallasnews.com/news/nationworld/mexico/20120116-for-most-mexicans-the-digital-age-is-still-out-of-reach.ece (accessed 16 August 2016).

Wilkins, J. (2005) Eating right here: Moving from food consumer to food citizen. *Agriculture and Human Values*, 22(3): 269–273.

Yuku Kafe (2016) Specialty Coffee Mixteca Facebook page. Available at www.facebook.com/Specialty.Coffee.Mixteca/ (accessed 16 June 2016).

5 Political consumers as digital food activists?

The role of food in the digitalization of political consumption

Katharina Witterhold

Introduction

Scholarly interest in political, ethical, and critical consumption has increased considerably over the last decade (Baringhorst, 2007, 2015; Gallego, 2007; Lamla, 2013; Micheletti *et al.*, 2003; Stolle and Micheletti, 2003, 2013 Stolle *et al.*, 2005; Yates, 2011). But despite growing research on non-institutional forms of political participation like food activism (Counihan and Siniscalchi, 2014), political consumerism[1] (Stolle and Micheletti, 2013), and digital citizenship (Bennett, 2008; Swigger, 2013), these literatures rarely relate to each other. This raises a challenge when attempting to understand how different forms of non-institutional political participation are interconnected. For example, it is unclear whether and how practices such as signing an e-petition, purchasing organic food, and taking part in local urban farming initiatives interrelate. To explore these non-institutional forms of political participation in relation to one another and to more conventional forms of political participation like voting, Van Deth (2014) has suggested that these forms of participation, which take place in the private or semi-public sphere, should be conceptualized as lifestyle politics (Giddens, 1991). This concept carries methodological consequences as lifestyle politics entails practices embedded in everyday life. These practices, which take place in mundane spaces like the home, shopping centres, and public transportation, are therefore difficult to study at times.

This chapter reports on findings from the empirical study 'Consumer Netizens' carried out at the University of Siegen, Germany, between October 2011 and August 2015. The study examined the shift towards lifestyle-related forms of political participation. In particular, the study focused on the use of social web applications for political consumption. A research design comprised of qualitative, quantitative, and mixed-methods approaches was developed, including data collection through participant diaries (see section 'Research design'). Drawing on these diaries, the chapter highlights the role that food plays in the digitalization of political participation. The chapter suggests that daily food consumption practices in the private sphere of everyday life are directly interwoven with (semi-)public social media practices that address political concerns as well as questions of self-identity.

A good share of what is commonly understood as political consumption involves purchasing or boycotting certain products and services for political or ethical reasons (Micheletti *et al.*, 2005). While food played an important role for all the political consumers who participated in the study 'Consumer Netizens', the ways in which food became politicized and integrated in digital and in offline contexts varied considerably (Witterhold, 2015). Recent anthropological research has emphasized the meaning of food as a political matter. Counihan and Siniscalchi define 'food activism' as 'the efforts by people to change the food system across the globe by modifying the way they produce, distribute, and/or consume food' (2014: 2). 'Food activism' becomes a strong symbol for other political issues in a society and a starting point for social movements which can be as disparate as the Slow Food Movement, Fair Trade or community-supported agriculture (ibid.: 4). The authors leave open the question of whether food activism can be regarded as a political movement explicitly: 'We include in food activism people's discourses and actions to make the food system or parts of it more democratic, sustainable, healthy, ethical, culturally appropriate, and better in quality' (ibid.: 6). Individual, collective, planned, spontaneous, local, global, one-off or continuing activities can be included here. Inclusive as this definition is, it raises the question of where to draw a line between activism and non-activism when nearly every food-related practice can be understood as having an impact on the food system. For example, can the use of a Payback card,[2] which collects data on individuals' shopping practices, be considered food activism? A participant in our study used her Payback card when she bought vegan products. Though she knew that Payback is about collecting consumer information (to improve individualized advertisement), she was convinced that her consumer data could alter the statistics and thereby boost the production of vegan food. While this explanation may make sense for the research participant, from another perspective it seems to be more likely that Payback would make use of the data to alter consumer behaviour and not the other way around. So the question remains, when is it helpful to talk about 'activism' or 'political participation'? To leave this decision to citizens themselves could, as van Deth argues, end in a subjectified concept of political action (2014: 349). A fruitful starting point would be to bring together the participants' different accounts of lifestyle politics and develop a common conceptual frame of political action.

The intersection of political consumerism and online netizenship: consumer netizens

Since the rise of the so-called new social movements (Offe, 1985), the meaning of the term 'political' has become more and more contentious (Dalton, 2008; van Deth, 2014). Social scientists disagree about how to define what counts as political participation. The main focus of this discussion is political praxis, which is situated in the sphere of everyday life, and where, through lifestyle, discourse, art, or consumption, political aims are pursued outside the conventional realm of policy. Giddens (1991) interpreted the politicization of everyday life as a consequence of the dynamic modernization of society (particularly with respect to knowledge about oneself and the consequences of individual acts); while this politicization has occurred, conventional forms of political participation, such as party membership, have declined (Crouch, 2008; Putnam, 1995). Moreover, from a theoretical position, the concept of a politics of everyday life has been criticized for devaluing the term political: If everything is political, nothing would be political any longer (van Deth, 2014: 350). This on-going dispute has been revitalized by emerging forms of web-based participation. Again, there exist contradictory assumptions on how to assess the political content of these practices: do they represent mere 'clicktivism' (Morozov, 2009) or the rise of a new participatory culture, where consumers become active 'produsers' of media content (Bruns, 2008)?

Similar concerns have been raised with regard to political consumption. As a 'consumer citizen', the person directs his or her consumption practices towards the rights of third parties: succeeding generations, workers, animals, plants, and the planet. However, the motivations of political consumers have been contested (Gilg *et al.*, 2005; for an overview of the discussion, see Thogerson, 2011). Are these consumers really motivated by the common good, or are they using their economic power for distinction and status (Griskevicius *et al.*, 2010)? On the other hand, several social scientists view consumers' expanding orientation beyond their so-called needs[3] as a new form of political participation (Micheletti, 2003; Baringhorst, 2010). This process, which has been described by Lamla (2013) as 'consumer democracy', could, if expanded, become relevant on a global scale.

New forms of online political participation have more in common with political consumption when a closer look is taken at the underlying structures of action. Parallels between online political participation and offline political consumption are observed in their situatedness in the sphere of everyday life, where this kind of political action mostly takes place, and in their reference to knowledge gained from practical experience rather than expert knowledge (Waldschmidt, *et al.*, 2009).[4] Additionally, when it comes to the question of values, both emphasize personal autonomy and global solidarity (Bennett and Segerberg, 2013). These commonalities could be interpreted as hints at a general transformation of citizenship. Referring to the work of L. Bennett (2008), Baringhorst (2007, 2016) argues that the shift from lifestyle-related practices towards a new understanding of citizenship and policy arises from the merging of political consumption practices and social web use in everyday life (Baringhorst, 2012). In relation to political protest becoming part of everyday life, Baringhorst writes about 'new opportunity structures'. This addresses processes of transformation in the political, economic, and media fields, which 'link the role of the citizen to the status of the consumer' (Baringhorst, 2010: 12). In the political realm, Baringhorst argues that the emergence of new structures of governance should be taken into account. These structures could be characterized through the growing participation of different actors (nation-state representatives, European, international, and civil actors) in problem-solving. In these processes, she expects non-governmental organizations to increase in significance, while liability and responsibility remain unclear (ibid.: 15).

Since the 1990s, analyses of new information and communication technologies have suggested that the rise of the Internet and especially the spread of social web technology would enable consumers to connect to each other, share information about products and services, and collect and compare information about companies and their practices (Scammell, 2000). According to theories of networked participation and networked public spheres, social media (like social networking sites, wikis, and weblogs) are supposed to enable and motivate consumers to become active 'produsers' of media content (Bruns, 2008). With the term 'produser', i.e. the hybrid of 'produce' and 'user', Bruns refers to the well-known term of the 'prosumer' that Toffler (1983) introduced to describe his observation of an on-going blurring of the boundaries between the producer and the consumer. However, it was Bruns who realized that this term also describes modes of self-production or collaborative production in digital contexts. However, in research operating at the intersection of online and offline, the term 'prosumer' still holds analytical relevance. Here, the term 'prosumer' might be more applicable as an umbrella term, because Bruns' 'produser', with its dependence on social media, seems to be a specific form of prosuming. According to Toffler, Bruns (2008) proposes a distinction between two types of prosuming: the professional and the producing consumer. The 'professional consumer' defines a consumer who has gained some expertise in one field of consumption. This expertise allows the consumer to give the producer highly informed feedback about products and advice for further developments. But this form of consumer participation in the production process does not alter the communicative hierarchy between producer and consumer. In Bruns' view, only through the digitalization and the development of Web 2.0 is the consumer enabled to destabilize these hierarchies. This is what Bruns defines as the producing consumer or the 'produser'. It is important to note that Toffler and Bruns refer to different sorts of consumer goods. While Toffler located the prosumer in relation to the production of consumer goods and services, Bruns' produser collaborates in the production of digital – or more precisely, information – goods. Digital goods are considerably different from consumables, because they do not lose value when used; indeed, the more content is 'prodused', the more it is valued. This differentiation becomes clearer when modes of food- related political consumption are viewed as prosuming or produsing. When a prosumer becomes a produser by sharing her knowledge and motivation on the Internet, this could be seen as an attempt to 'become majority', one of the basic requirements for political action (Marchart, 2013: 301).

These complex relationships entailed in the intersection of personal lifestyle and the political have led scholars to turn to practice theory (Sahakian and Wilhite, 2014) in order to understand the everyday practices of consumption (see also Jäger-Erben and Offenberger, 2014). In their research on sustainable consumption, Sahakian and Wilhite's (2014) praxeological perspective takes into account the relations between products, bodies, and social fields. Practices are the empirical point of access for such a perspective: they connect products, subjects, knowledge, and social structures (ibid.: 26–27). Practices can also be considered as routines; i.e. certain modes of behaviour in similar situations – not identically repeated, but consisting of joint elements (see Reckwitz, 2005: 7). Social practices should not be mistaken as 'automatism', but as a practice expression[5] of (collective) social reason[6] (ibid.: 8). There is no coherent research paradigm for practice theory; however, researchers share a common understanding of defining social practices, spatially, temporally, and materially embedded in a situation or context, as the central analytic unit (ibid.). A practice theory perspective is especially helpful in the analyses of actor-actant-interactions[7] too, whereby the practices are not only understood as acts, but as expressions of both interpretation and appropriation (ibid.: 285).

The research project 'Consumer Netizens': methodological approach, and an overview of the results

Research design

The research project 'Consumer Netizens' studied the attitudes, interpretations, and practices of political consumers online and offline. The project's aim was to identify new, web-based forms of political participation and related (new) self-definitions of citizenship. During the first stage of data collection, participants were asked to keep participation diaries. Following Markham and Couldry's (2007) media diary method, participants were asked to document their everyday life, focusing on consumption, media use, and political participation for at least eight weeks. Participants were recruited by calls for participation posted on several websites (e.g. www.karmakonsum.de accessed 28 April 2017 or www.utopia.de accessed 28 April 2017) which had been identified in a previous study (see Yang and Baringhorst, 2014). First, potential participants had to fill out a short online survey, containing questions about their activities as political consumers online and offline, age, education, and federal state. Respondents who resided in Nordrhein-Westfalen, Berlin, Bremen, and Hessen were included in the study, to facilitate in-person sessions with the researchers: first, on diary writing instruction, and then for a post-diary writing interview. In conceptualising the diary method, the research team considered the extent to which participants should be instructed about the content or forms of their diaries. In the end, participants were not given specific instructions, to allow them to choose the content most relevant to their lives and understandings of citizenship. Instead, the participants were invited to keep daily records about those aspects of their everyday life, including thoughts, which were connected, in their view, to political participation, consumption, and media use. The format was flexible, to ensure that participants felt comfortable during the writing process, and participants could decide if they preferred to write a diary entry via email and receive feedback from a researcher every day, or if they preferred to write for eight weeks, using pencil and paper, without feedback.[8]

At the second stage of data collection, semi-structured interviews were conducted. The first part of the interview contained questions concerning the life history of each participant and how they became interested in political consumption, as well as questions about passages from the diary that were unclear or that included content of particular interest for the study. The second, more structured part of the interview, included questions about the participants' general political and consumer attitudes, e.g. what they thought about the consumer society, or who (government, NGOs, consumers) had responsibility for addressing problems that consumers face.

The analysis investigated not only political participation practices, but also attitudes, self-definitions as citizens, collective interpretations of citizenship, and representations of these new types of citizens. In addition to the diary and the semi-structured interviews, two group discussions were conducted. Based on a multi-stage interpretation process, types of consumer netizens were constructed. Finally, the typology was tested through a quantitative survey for which an online panel was used.

Participants

As mentioned above, potential participants first took part in an online survey. Afterwards, researchers contacted potential participants (who resided in the study areas) via email to obtain contact details and to arrange a first telephone conversation. The potential participants received detailed information about the research project and what would be asked of them, and about the data protection regulations applied.[9] Then, face-to-face meetings were arranged. These took place near or in the homes of participants. Every contact of a research team member[10] with participants was noted in a protocol, including information about what had been said, alongside subjective impressions, to help account for personal assessments and biases. These face-to-face meetings were conducted to develop a climate of trust and obligation and prevent attrition. In total, 26 participants (12 female and 14 male) completed the diary and were interviewed; the participants' age range was 18–55 years (with only one person older than 45 years), and 17 had an academic background, with 8 having a university degree. Less than half of the participants took part in the group discussion. This was due to the difficulty of arranging a discussion in Siegen on a date that would suit all participants, rather than because of the participants' unwillingness to participate. All things considered, the research team was in contact with the participants for over two years.

Analyses

The diary method produced very heterogeneous data. The diaries ranged from tabulated buying and website protocols (where participants documented which websites they visited) to creative picture portfolios with aesthetic aspirations. Due to the heterogeneity of data, the analysis required different interpretative methods. Those diaries that focused on daily consumption and media practices were coded using the qualitative data analysis software MAXQDA,[11] based on the principles of grounded theory; later, the codes were developed into categories on axes such as online–offline, private–public, and exit–voice (Hirschman, 1974).[12] Other diaries called for techniques such as documentary image analyses (Bohnsack *et al.*, 2013). For narrative passages, the sequential analysis of objective hermeneutics (Övermann, 2000) was applied. Because the diaries contained many hyperlinks to webpages and even personal social web pages, they challenged a strict sequential analysis, and required an analysis of, in part, very intimate data. Though it required much more time, the hyperlink structure of the diaries was followed when it was necessary to understand the sequence's meaning. However, hyperlinks or screenshot citations that could compromise the participants' privacy were omitted.

The praxeological analysis of the diaries suggests the existence of different types of 'consumer netizens' and citizen practices. Common to all diaries was the participants' focus on consumer political topics and questions, but the ways in which they used the Internet and especially the social web to pursue their interests differed significantly. These differences in online practices corresponded to differences in offline practices, and, in particular, the ways in which the participants included food politics in their everyday life.

In the analysis, the focus was on practices of reproduction (routines) and the potential transformation of routines. The aim was to examine if and how the use of social web applications altered or extended the political practices of the participants. Following the principles of grounded theory (Strübing, 2004), practices were analysed in the core dimensions of exit–voice, private–public, and offline–online. In the course of the analyses, the analytic scheme became more detailed. It became more and more obvious that different fields of action (the economic, the cultural, and the political) were linked to different political consumption types. While consumer surveys tend to exclusively concentrate on the economic field, the diaries revealed the significance of the cultural and political dimensions. At least four types of practices were identified: (1) Green Buycott; (2) Expressive Lifestyle; (3) Produsage; and (4) Integrative Prosuming. Although general assumptions about the broader consumer public cannot be made, it is important to note that women participants were most represented in types 1 (Green Buycott) and 2 (Expressive Lifestyle), while the third type, 'Produsage', was discerned in the diaries of men participants only. The fourth type, 'Integrative Prosuming', which differed widely from the other types, represented only two diaries (one written by a woman, the other by a man).

Overview results

Type 1: Green Buycott

The first type can be characterized by political consumption practices which remain in the private sphere of everyday life, where they centre on food as well as on environmentally friendly behaviour, such as travelling by bike instead of driving. Internet use is directed towards the facilitation of the economic and social organization of everyday life. Publishing content or signing e-petitions are not part of the praxis. Social networks are used, but only to help organize existing social relations with friends offline. The politicization of food translates into the pleasurable means of boycott. Organic food replaces non-organic food and, where circumstances allow, gardening and farming practices can be observed. Participants whose practices belonged to this type were all women who, through an academic education, had achieved expert knowledge on consumer affairs to which they could refer to develop criteria for their political consumption. Bodily aspects such as health and appearance played a major role in shaping their daily consumption routines. Besides food, products which directly touch the body, like cosmetics, were politicized. It seems that health concerns – at least for that type – can enhance or initiate acts of political or at least sustainable consumption:

> I: In your diary, you had mentioned another aspect you haven't referred to yet, actually such cosmetic stuff, all these…
>
> S: Yes, there, I really buy sustainable. And I would like to maintain that. But I think, there, that will rather prevail in the long run, just because of all those allergies and skin diseases, people suffer from today. I can see it with my mom, who somehow sells more and more cosmetics in the pharmacy, stuff without perfume, which are sold now. And I also buy sustainable detergents and such stuff without any additives … such things; I pay very close attention to.
>
> (Interview with Lara, paragraph 99–102)

Type 2: Expressive Lifestyle

The second type, called Expressive Lifestyle, also has a strong tendency towards acts of buycott. With the exception of one diary, all diaries belonging to this type were written by women. Similar to Green Buycott, the focus is on buying organic food, seasonal, and regional products. In contrast to type 1, practices are orientated towards ethical concerns such as the rejection of animal products. This self-imposed obligation shapes everyday consumption and social media practices to a great degree. Besides animal rights, the rights of children and workers are also considered important criteria for consumption decisions. The use of the Internet is guided by the desire to raise awareness for these ethical concerns, to show how consumption that follows these criteria can be possible, and to develop a kind of ideal online social self, which participants try to realize in their everyday lives offline. Social media practices centre on one social network (Facebook) for the mutual exchange of information about products and experiences, and also approval (or disapproval when faced with 'old friends' who do not share this lifestyle). New consumption practices tend to take the form of experiments and self-imposed challenges like: Can I live vegan for six months? Can I consume 100 per cent ecologically and socially-correct products for one week?[13] The consumer is thus open to trying new products, new shops, and new tools for self-assessment. In contrast to type 1, participants engage in public practices such as signing e-petitions and publishing their own content – not only comments and links, but also photographs of self-prepared (vegan) dishes. Giving voice to one's practice online is very important, not only in addressing others via a personal social web page, but also in searching for contact with like-minded people all over the world. Although these participants sign several e-petitions per day, they describe themselves as rather non-political when asked about their conventional modes of participation offline. Though all of them vote, they seem not to perceive themselves as politically active citizens.

Type 3: Produsage

As mentioned above, practices of this type, Produsage, were represented in the diaries of male participants only and were focused almost completely on the Internet. A daily routine of political consumption could not be observed. These practices are directed at the political and cultural dimensions: the search, co-production, and distribution of consumer-related knowledge, new technologies of production, but also new modes of political participation. They actively participate in a variety of social networks online, which are used strategically depending on their respective concerns. They scan the Internet for information, but in contrast to type 2, they are less interested in the aesthetic dimension than in the technical and innovative aspects of a new product, invention, or software. Participants take part in projects like open source and crowd-funding, and campaign for new modes of political participation like liquid democracy.[14] Though political consumption practices such as buying organic food or Fair Trade clothes are not frequently part of this type, in contrast to types 1 and 2, the social web practices of type 3, like the forwarding of information on Facebook and Twitter, connect them to users in types 2 and 4 (described in the next section).

Type 4: Integrative Prosuming

Integrative Prosuming can be seen as connective work, in which social web practices are linked with political discussions online as well as with practical problems emerging from everyday life offline. These practices take place in the public and private spheres. Like the first two types, membership of parties or organizations is rejected, but informal connections to organizations, for example, Greenpeace, exist, and participants support these organizations' causes from time to time. Like type 1, an established routine in consumption practices offline can be observed, but like type 2, reflecting upon these practices offline and online is a crucial part. Therefore, the exchange of experiences and information with others online is emphasized. When publishing the results of web searches, the information is framed in a way that makes it more consumable for others. For example, one participant uses two Facebook accounts. The first account is under her own name, and includes Facebook friends she has met offline. For the second account, she has created the virtual character of Waldtraud,[15] a female tree. Waldtraud's background story concerns humans' loss of interest in and understanding of nature. This is why 'Waldtraud' decided to log into Facebook, where she regularly tells stories about the insects living in and on her bark, how important she and her 'relatives' are as wind-protectors, and sometimes about her grief, when 'family members' are felled. Though the invention of Waldtraud may be a relatively extreme example, it serves well to show the creativeness and the efforts of type 4 consumers. Offline they actively take part in local citizens' initiatives like planting trees or teaching school children how to build hotels for insects. Animal rights are an important issue, and they reject the consumption of meat (but not milk-based products). However, in contrast to type 2, whose engagement online is overwhelmingly concerned with food and animal-related topics, animal rights are only one of many interests, which also include environmental concerns, the fight for free speech, and tighter regulations for large corporations like Monsanto.

The role of food for consumer netizens

Food-related practices are at the core of at least two political consumption styles: Green Buycott and the Expressive Lifestyle. The praxis of Produsage can be excluded, because food-related routines cannot be identified either offline or online for this type. As for the fourth type (Integrative Prosuming), the issue of food is important, but it is framed by a more general environmental concern. Most interesting is the comparison of type 1 and type 2, because their politicization is connected directly to the issue of food. The following discussion examines two contrasting case examples of Green Buycott and Expressive Lifestyle.

Annika Seifert[16] (Green Buycott)

> Em, what I can say, or what I'm almost sure about is that the issue of sustainability became important for me through the issue of food. And that the issue of food and delightful consumption and so on always used to be important in my family. Not regarding sustainability, however, but really regarding pleasure and so on. And that my parents themselves used to have a garden and to grow their own food, for sure that is something that led me to appreciate that in some way.
>
> (Interview with Annika, paragraph 17)

Annika was first exposed to political consumption as a student. She enrolled in a course on world food politics, facing questions like food justice and the consequences of western consumption styles. This marked the beginning of her political consumption practices that later evolved to include the scientific aspects of food politics. When the research project started, she had just finished her studies and was about to get married. In her diary she described the preparations for her wedding that she entitled 'my green wedding'. However, most of her plans could not be realized due to the economically well off but rather conventionally orientated family of her fiancé. Having lived with him, she was used to his parents' and his friends' attitude towards consumption. Instead of becoming absorbed in endless discussions about what to eat, she tried to convince them to consume more consciously through presents of high quality organic olive oil or Fair Trade organic chocolate, following the conviction: Good taste asserts itself. In the interview she emphasized that it is every consumer's free decision what to eat and what to buy, though later, referring to the question of what should be done with respect to consumer policy, she stated that only those who possess specialized knowledge, which cannot be expected from the majority of consumers, can make good consumption decisions. The solution she therefore proposed was to extend the cooperation between political and consumer experts.

Jana Peters (Expressive Lifestyle)

Every day I get up, I tell myself: 'Do not buy palm oil, Jana!' but I might not then always manage to do that. These are perhaps those conscious things, I think about. Will I have to go shopping today and if, can I just afford to buy bio-fair or will it 'only' be vegan, because vegan and bio and fair don't mean the same. Of course, this would be the product of my dreams, but it isn't completely workable and it is a question of money, too. Well, I, if you have more money, it's definitely easier.

(Interview with Jana, paragraph 63)

Jana's case involved different and partly opposing dynamics at work. Jana's politicization started when she consulted a nutritionist. She decided to see a nutritionist partly because of her skin allergies, but also because she suffered from being overweight. This visit to the nutritionist took place shortly after Jana moved out of her parental home. However, it marked the point where Jana began to reflect on her eating habits. The nutritionist's rather vague suggestion about a potential interrelation between food intolerance and animal products made her turn to interest groups on Facebook. Their advice and codes of conduct served as an important anchor for reorganizing her consumption habits offline. But the integration of opposing criteria for adequate political food challenged her entire consumption style. Until a new consumption routine could be developed, she engaged in what might be characterized as green compulsive buying, mixed with feelings of guilt when she could not achieve her green consumption goals. In the course of the project Jana became able to mediate between the norms of different consumer action groups and her own successively discovered preferences. Gaining more and more self-confidence in what she did – e.g. learning the meanings of different labels for bread – she communicated her insights, her thoughts, feelings, and experiences via her profile on Facebook. On Facebook, Jana was connected to many different groups like foodwatch, the 'Vegetarierbund' (Vegetarian Union), and PETA (People for the Ethical Treatment of Animals), and also to people from all over the world, who shared her views and attitudes towards consumption (most of her posts were in English). Her profile on Facebook served as a kind of diary through which she documented her way to a 100 per cent 'correct' lifestyle.

At the start of the study and in contrast to Annika, Jana stated that she was not interested in politics. She did not even vote. This attitude had changed over the course of the two years: She developed a political identity as a consumer and a citizen, using all means at hand to pursue her goals. In the end, she was not only a regular buycotter of certain products, which she actively promoted via her Facebook profile, but also signed petitions, took part in the 'Bundestag' vote and took her first steps into the sphere of participation offline, going to demonstrations and to action group meetings. Though her politicization started by relating weight and health concerns to organic and Fair Trade food, it developed into a wider commitment to human and animal rights.

Comparing the case studies

In both Annika's and Jana's cases, becoming a political consumer was strongly related to food. Both suffered from allergies and were thus particularly attentive to the connection between food and health. Annika, who has gained expert knowledge due to her academic education, translated that into a politically adequate shopping cart. The dominant dimension of action here was the economic one.[17] Beyond that, she did not share her expert knowledge in discursive political consumption, i.e. on social media forums for political consumption, nor did she promote her own lifestyle as an example of successful and enjoyable political consumption. Referring to that, she said she found social web-based communication senseless:

> Exactly. Well, the exchange online, I do not think that it is incredibly constructive in principle and it isn't very helpful, too. And this is why I do not take part in a debate about, I don't know, is E10 actually better or whatever, I do not personally have any added value, when I'm participating there.
>
> (Interview with Annika, paragraph 108)

Apart from food, Annika did not question her pleasure-orientated lifestyle: Seemingly unhealthy, low-quality products were replaced by organic ones. Here, the women's social environment played a significant role. While Jana lived on her own, Annika had just married into a family who showed a strong orientation towards conservative values (expressed by the desire to maintain the status quo in personal lifestyle as well as in the organization of society, e.g. in modes of food production). Annika only succeeded in asserting her own values when she was able to relate sustainability to the status-orientated aspects of an improved lifestyle. She found support in her working environment – she and her colleagues develop educational units to teach sustainability – but it seems to be quite difficult to translate expert knowledge into practical knowledge, particularly when each attempt was preceded by a family discussion.

By contrast, Jana's daily search for, and validation of, information were followed by publishing the most interesting or apparently important news online on her social media page. Thereby, her communication style was highly emotional and expressive, possibly because she was establishing a connection between herself and the news.

> But of course, that [political consumption] is a power, and this is, I think, there, I don't have to be that politically super-informed, because I can start at a small circle and many pennies make a dollar ... And there is so much and I always love to post my maxim: 'Do a good deed every day' and 'Here, look at this, sign it, it's a good thing' and then you have brought about something already. And then, 100,000 of other people do that, too, yes, and, through this, some politician gets handed over some scrap of paper. But if people don't do that, the politician won't notice anything. And that's simply little steps, which, in my opinion, can be so easily integrated into everyday life and this is why I cannot understand why everybody is not doing it.
>
> (Interview with Jana, paragraph 61)

Conclusion

This study explored the digitalization of lifestyle politics using case studies of political consumption. The analysis focused on qualitative data collected through participant diaries and semi-structured interviews to examine how food activism, political consumption, and digital citizenship are connected. The case studies demonstrated the roles that food comes to play in the everyday practices of political consumers, first, alongside the impact that social web use has on a political consumer's food-related praxis. The participants' political consumption practices were characterized as political prosuming, in parallel with cultural studies approaches to consumption and political participation.

As the analysis shows, many political consumption practices are related to food. But these practices cannot be reduced to only the buycott or boycott of certain brands or products. The insights provided in the participants' diaries reveal an often creative form of market participation in which economic concerns are not foremost and which bypasses marked structures. This observation contrasts with the familiar concept of political consumerism introduced by Micheletti (Stolle *et al.*, 2005). There, the term is used to describe practices directed towards altering market structures. But, in this study, consumers who actively responsibilized themselves to counteract globalized mass production of food – through, for example, the boycott of meat, canned food or exotic fruits – also created alternatives to mass marketed foods, through growing vegetables, making preserves or, like Annika, contributing to the sustainability of old plant species by cultivating them and multiplying their seeds. Therefore, these consumers are better defined as prosumers. Here, the term prosumer can be applied in the classical meaning that Toffler (1983) intended when he described prosuming as production for self-use, from self-cooked dinners to self-help groups.

Toffler argued that formerly depreciated consumer work – mostly the work done by housewives – gains recognition when more and more people turn to the self-production of consumer goods (cf. ibid.: 388–390). But the turn to a prosumer's ethics not only implies spending more time in prosuming practices, but also implies the (re-)appropriation of knowledge. In the pre-industrial era, prosuming was an incremental part of everyday life, but was forgotten or delegated in the course of industrialization (ibid.). Now, for political consumers, this prosumer knowledge has become relevant again to regain, at least in part, independence from market structures while reinstating perhaps traditional and gendered forms of care work, such as food growing, preserving, and cooking. But should political prosuming practices be integrated into the concept of political consumption?

Though market-centric practices like product boycotts form an important part of political consumption, participants' diaries document a wide range of non-market-centric activities. These include not only the prosuming of food, but also preparing gifts through handicraft, repairing household items, manufacturing one's own furniture, exchanging clothes,[18] and writing computer programs for one's own use. All these practices refer to and rely on prosumer information and knowledge. Defining political consumption only as an economic, market-centric activity would mean disregarding much of the labour and creativity in which political consumers engage. To consider prosuming as a part of political consumption would modify notions of political consumption as a cheap and time-saving form of political participation. Therefore, I would like to suggest that the concept of political consumption should be broadened to include non-market-centric practices like prosuming.[19]

Why did food serve as a starting point for the participants' political consumption? The roles food comes to play may best be explained through the connection between food, personal health, and the public good – a connection stronger than, for example, the one between clothes, personal benefit, and the public good. First, food is part of a broader and on-going discourse about health and self-control, which aims to impose new modes of self-regulation and self-responsibility (see Keupp, 2006; Schneider and Davis, 2010). This discourse strengthens the link between food and health and raises awareness of food-related information. Second, food can be grown and prepared by consumers, providing a visible and tangible proof of time and labour investment. While the impact of individual political participation, especially in the mass food retail market, can rarely be observed, the prosuming of food enables an immediate experience of efficacy. In the case of Expressive Lifestyle prosumers, such as Jana, social web use helps to connect political and consumer identity, which leads to an individualization of participation, a passion for certain values (especially solidarity, justice, and animal rights), a rejection of membership of political parties or organizations, and a politicization which encompasses one's lifestyle as a whole. This goes along with a general rejection of formal membership in traditional political organizations. Jana exemplifies an emerging form of membership – one more loosely defined and centred on social media. Accordingly, her political consumption can be understood as an exploratory movement, one characteristic being the 'relation' to physical experience. In this respect, food serves as an important medium between the outside world and the body, potentially combining personal and political values when personal eating habits are articulated politically on the social web. By enabling people to share personal experiences, including descriptions of embodied experiences, the social web promotes the discursiveness of everyday life, where practices formerly enacted more or less unconsciously become important issues on personal social web pages. Here, a new kind of solidarity develops which originates in sharing personal experiences that might help others redirect their consumption.

The way political consumers inform themselves and talk about their information and knowledge work[20] indicates a transformation in the ways in which the self is perceived within the hierarchy of companies and consumers, as well as within the hierarchy of citizens and political institutions. At the same time, some prosumers tend towards a professionalization of political consumption, as in Annika's case, which seems in opposition to collective problem solving and common discourse. If a prosumer is guided by political or at least ethical reasons, would she not try to share her knowledge about prosumption with others? Web 2.0 provides an infrastructure for communication and collaboration that facilitates the sharing of knowledge and experience. In spite of this, many of the study's participants did not use the Internet to connect to other consumers. Annika stated that consumer interaction on the Web 2.0 did not offer her any added value. To explain the different attitudes towards consumer collaboration online, Bruns' (2008) differentiation between the 'professional prosumer' and the 'producing prosumer' is useful. According to Bruns, the professional prosumer is a consumer who has expert knowledge which allows her to make educated decisions. The understanding of political consumption should not be restricted to political participation, but acknowledge that it is an arena for economic and political experts, too, where new products, laws, and services emerge. Furthermore, it is an emerging field of knowledge which produces new frontiers to define experts and layperson. Therefore, the reluctance of consumer experts, like Annika, to collaborate with other consumers might be due to the more general problem of collaboration between experts and laypersons (Bourdieu, 2013: 98–99). Yet Bruns' reasoning reproduces an online/offline dichotomy. Offline prosumers, too, participate in practices of sharing knowledge within the family, with friends, and with colleagues. Additionally, 'produsers' often act as 'prosumers' when they cook 'ethical' dishes – and sometimes as both, when they photograph these foods and upload the photographs on Facebook. (In general, participants used Facebook. Twitter, Geraspora and online discussion groups were used only by a small number of male participants.)

The analyses above show both the professional and the producing prosumer as types of consumer citizens. Professional political consumers like Annika concentrate on the market economic and private dimension, where political consumption is practised through buying organic food or partly growing it. Motives underlying these practices include personal health, a relatively abstract orientation towards justice, and the conviction that these products are of better quality and therefore signal a better lifestyle. There is no general attempt to convince others to join this lifestyle, because the ordinary consumer is perceived as lacking the background knowledge needed to make the right consumer decisions. Annika herself has acquired such knowledge through her academic education. Thus, she emphasizes the cooperation between consumer experts (like herself) and political experts in developing directives for consumers and companies to enable more sustainable lifestyles. This is unlike the producing political consumer, Jana, who is convinced that it is her and every consumer's duty to reflect upon consumption and its consequences. Perhaps it is for that reason – the perception of a general responsibility – that Jana's political consumption is not only focused on organic food, but also on Fair Trade and animal protection. Because she believes in the responsibility and the ability of consumers to change consumption habits, she uses her personal Facebook profile to make public her own consumption lifestyle – what she buys, eats, wears, and why. In this way she produces an online identity helping her to direct further consumer decisions and motivate others by role-modelling consumption decisions.

The study has explored connections between food activism, political consumption, and digital citizenship. Alongside health concerns, the underlying motive for prosuming, as expressed by participants, is a desire to re-establish agency as a consumer. In the realm of prosuming, food has a particular resonance as it is incorporated into the body. However, it is remarkable that those political consumers whose practices centred on food (Green Buycott and Expressive Lifestyle) did not voice support for expanding consumer rights regarding food, e.g. through disclosure of ingredients. Those consumers who campaigned most for consumer rights were the least engaged in food politics (types 3 and 4 – Produsage and Integrative Prosuming). This suggests that food-related political consumption currently takes place primarily in the economic and cultural spheres, and is still tangential within the political sphere. If an expanding consumer movement is to develop, it is essential to bring together practice and lifestyle-bound food activism with the more conventional realm of consumer politics.

Notes

1 The term 'political consumerism' stresses the collective dimension of consumer engagement. In this chapter, the more inclusive term 'political consumption' will be used to take more individualistic forms of consumer engagement into account. Another issue is that 'political consumerism' is an ideology of political engagement through shopping, while 'political consumption' can also turn against the consumerist angle.
2 See www.payback.de (accessed 27 February 2017). Payback is a subsidiary company of The Loyalty Partners Group (www.loyaltypartner.com/) (accessed 27 February 2017). Consumers can sign up for Payback with their name and address and receive a card through which they can collect points for purchases. These points can be used to receive discounts on certain products on the Payback website.
3 Offe (198) argues that different kinds of needs have to be distinguished, but does not refer to sociological perspectives where basic and luxury needs are distinguished. Instead, Offe questions how much of consumers' 'need' is forced on them by external

conditions. For example, the need to have a car may be caused by the need to drive to work and the absence of public transport.

4 Waldschmidt *et al.* (2009) studied discursive knowledge production using data from a pilot project website about bioethics. Contrary to their expectations, one result was that participants contributed to the discussion about bioethics by referring mainly to everyday knowledge instead of expert knowledge (ibid.: 167).

5 In German, the term 'Ausdrucksgestalt' is used.

6 Bourdieu (1982) used the term 'sens practique'.

7 This specific combination of actor-network theory and practice theory is not new. In Germany, it has its roots in the sociology of technology (Rammert, 2003). However, it has not yet been applied in the study of online practices. As Reckwitz (2005) states, practices – even those of browsing and clicking – are still physically located in bodies which remain offline. Following this, the study of online activism cannot dispense with the offline sphere when using a practice theory approach. Though the online sphere cannot be understood as a closed-up cultural space (see here also Lamla, 2010, actor-network theory helps one to be aware of the potential of technology and its interfaces to shape a routine that has been digitalized.

8 For some participants it was quite difficult to follow the vague instruction 'just write about everything in your everyday life, what you think is linked to political consumption, political participation, and media use'. These participants sent their diary entries via email, to ask whether they wrote about the 'right' things. The attendant person from the research team replied to these emails, trying to motivate them, by writing back: 'That's very interesting, keep it up!'

9 All participants signed a declaration of consent that permitted the anonymized use of their data.

10 The members of the research team were Sigrid Baringhorst, Mundo Yang, and Katharina Witterhold. See http://blogs.uni-siegen.de/consumer-participation/ (accessed 16 October 2017).

11 MAXQDA is software developed by Udo Kuckartz, Professor of Methodology. It is based on the principles of grounded theory and enhances the coding and the comparative analyses of texts (see also Kuckartz, 2014).

12 Exit–voice refers to A. O. Hirschman's (1974) well-known concept of exit and voice as being two modes of showing dissatisfaction with a company or a (political) organization. There, the 'exit' can be understood as the convenient form of consumer behaviour, to buy or to stop buying, whereas voice refers to the political realm where citizens raise their voice to articulate their dissatisfaction. In Hirschman's approach, these two forms of coping with the decline of quality or justice can be used to describe changing forms of participation as well as to analyse the reciprocal relation between these forms.

13 It is interesting to note that political consumers would not use the term ethical or political to describe themselves.

14 Liquid democracy is used as a buzzword to point to modes of political participation using the opportunities the Internet provides to facilitate, for example, electronic voting, but also political discussion forums or forwarding information about current debates.

15 The name has been changed to protect privacy.

16 To maintain confidentiality, all participants' names have been changed.

17 Taking into account that Annika works in the field of consumer affairs, perhaps the cultural dimension should be subdivided into laymen and professional forms of knowledge production.

18 Participants used the online platform, www.kleiderkreisel.de (accessed 28 February 2017). But the exchange of clothes also takes place in the private and offline sphere: For example, Annika stated that she and her friends regularly organize clothes-exchange parties to get 'new' clothes while getting rid of 'old' ones.

19 Alternatively, one could follow Kotler's (1986) suggestion to overcome the boundaries between production and consumption, and understand prosuming as the standard activity in everyday life with occasional deviations to one pole (production) or the other (consumption). Consequently, political prosuming would be applied as an umbrella term, using political consumption only to describe those practices directed at the market.

20 It became necessary to distinguish between information and knowledge work in the context of social web use. Practices of information work concern searching and organizing data while knowledge work refers to the bodily aspect of integrating information into practices.

References

Baringhorst, S. (ed.) (2007) *Politik mit dem Einkaufswagen. Unternehmen und Konsumenten als Bürger in der globalen Mediengesellschaft* [The political shopping trolley: Companies and consumers in global media society]. Bielefeld: Transcript.

Baringhorst, S. (2010) Anti-Corporate Campaigning – neue mediale Gelegenheitsstrukturen unternehmenskritischen Protests [Anti-corporate campaigning: New media opportunity structures for political protest]. In S. Baringhorst, V. Kneip, A. März and J. Niesyto (eds) *Unternehmenskritische Kampagnen. Politischer Protest im Zeichen digitaler Kommunikation*. Wiesbaden: VS, Verlag für Sozialwissenschaften, pp. 9–31.

Baringhorst, S. (2012) Der Bürger als 'Produser'. Politische Beteiligung von Konsumentenbürgern im Social Web. [The citizen as 'produser': Consumer citizens' political participation on the social web]. In C. Lutz, D. Seitz and E. Rösch (eds) *Partizipation und Engagement im Netz: neue Chancen für Demokratie und Medienpädagogik*. München: kopaed, pp. 63–77.

Baringhorst, S. (2015) Konsum und Lebensstile als politische Praxis. Systematisierende und historisch kontextualisierende Annäherungen [Consumption and lifestyle as political praxis: Systematic and contextualizing approaches]. *Forschungsjournal Soziale Bewegungen*, 28(2): 17–27.

Baringhorst, S. (2016) Nachhaltigkeit durch politischen Konsum und Internetaktivismus – Neue Engagementformen zwischen postdemokratischer Partizipation und demokratischem Experimentalismus [Sustainability through political consumption and Internet activism: New forms of engagement between post-democratic participation and democratic experimentalism]. In G. Diendorfer and M. Welan (eds) *Demokratie als Beitrag zu einer nachhaltigen Gesellschaft. Herausforderungen, Potentiale und Reformansätze*. Vienna: Studienverlag, pp. 43–60.

Bennett, W. L. (2008) *Civic Life Online. Learning How Digital Media Can Engage Youth*. Cambridge, MA: MIT Press.

Bennett, W. L. and Segerberg, A. (2013) *The Logic of Connective Action: Digital Media and the Personalization of Contentious Politics*. Cambridge: Cambridge University Press.

Bohnsack, R., Nentwig-Gesemann, I. and Nohl, A. M. (eds) (2013) *Die dokumentarische Methode und ihre Forschungspraxis: Grundlagen qualitativer Sozialforschung* [The documentary method and its praxis of research]. Wiesbaden: Springer Verlag.

Bourdieu, P. (1982) *Die feinen Unterschiede. Kritik der gesellschaftlichen Urteilskraft*. Frankfurt am Main: Suhrkamp.

Bourdieu, P. (2013) *Politik. Schriften zur Politischen Ökonomie*, 2. Berlin: Suhrkamp.

Bruns, A. (2008) *Blogs, Wikipedia, Second Life, and Beyond: From Production to Produsage*. New York: Peter Lang.

Couldry, N., Livingstone, S. and Markham, T. (2007) *Media Consumption and Public Engagement: Beyond the Presumption of Attention*. New York: Palgrave.

Counihan, C. and Siniscalchi, V. (eds) (2014) *Food Activism: Agency, Democracy and Economy*. London: Bloomsbury.

Crouch, C. (2008) *Postdemokratie* [Post-democracy]. Bonn: Bundeszentrale für Politische Bildung.

Dalton, R. J. (2008) Citizenship norms and the expansion of political participation. *Political Studies*, 56(3): 76–98.

Gallego, A. (2007) Unequal political participation in Europe. *International Journal of Sociology*, 37(4): 10–25.

Giddens, A. (1991) *Modernity and Self-Identity: Self and Society in the Late Modern Age. The Emergence of Life-Politics*. Cambridge: Polity Press.

Gilg, A., Barr, S. and Ford, N. (2005) Green consumption or sustainable lifestyles? Identifying the sustainable consumer. *Futures*, 37(6); 481–504.

Griskevicius, V., Tybur, J. M. and Van den Bergh, B. (2010) Going green to be seen: Status, reputation, and conspicuous conservation. *Journal of Personality and Social Psychology*, 98(3): 392–404.

Hirschman, A. O. (1974) Exit, voice, and loyalty: Further reflections and a survey of recent contributions. *Social Science Information*, 13(1): 7–26.

Jäger-Erben, M. and Offenberger, U. (2014) A practice theory approach to sustainable consumption. *GAIA*, 23(1): 166–174.

Keupp, H. (2006) Gesundheitsförderung als Identitätsarbeit [Health promotion: An issue of active identity formation]. *Zeitschrift für qualitative Bildungs-, Beratungs- und Sozialforschung*, 7(2): 217–238.

Kotler, P. (1986) The prosumer movement: A new challenge for marketers. *Advances in Consumer Research*, 13: 510–513.

Kuckartz, U. (2014) *Qualitative Text Analysis. A Guide to Methods, Practice and Using Software*. London: Sage.

Lamla, J. (2010) Zugänge zur virtuellen Konsumwelt. Abgrenzungsprobleme und Revisionsstufen der Ethnographie [Accessing the virtual world of consumption.] In F. Heinzel *et al.* (eds) *Zugänge zur virtuellen Konsumwelt*. Wiesbaden: VS, Verlag für Sozialwissenschaften, pp. 127–139.

Lamla, J. (2013) *Verbraucherdemokratie. Politische Soziologie der Konsumgesellschaft* [Consumer democracy: Political sociology of the consumer society]. Berlin: Suhrkamp.

Marchart, O. (2011) Democracy and minimal politics: The political difference and its consequences. *South Atlantic Quarterly*, 110(4): 965–973.

Markham, T. and Couldry, N. (2007) Tracking the reflexivity of the (dis)engaged citizen. *Qualitative Inquiry*, 13(5): 675–695.

Micheletti, M. (2003) *Political Virtue and Shopping: Individuals, Consumerism, and Collective Action*. New York: Palgrave Macmillan.

Micheletti, M., Stolle, D., Nishikawa, L., and Wright, M. (eds) (2005) A case of discursive political consumerism: The Nike e-mail exchange. In *Proceedings of the 2nd International Seminar on Political Consumerism*, pp. 255–290.

Morozov, E. (2009) The brave new world of slacktivism. *Foreign Policy*, 19(5).

Offe, C. (1981) Ausdifferenzierung oder Integration — Bemerkungen über strategische Alternativen der Verbraucherpolitik [Differentiation or integration: Remarks on strategic alternatives to consumer policy]. *Journal of Consumer Policy*, 5(1–2): 119–133.

Offe, C. (1985) New social movements: Challenging the boundaries of institutional politics. *Social Research*, 52: 817–868.

Övermann, U. (2000) Die Methode der Fallrekonstruktion in der Grundlagenforschung sowie der klinischen und pädagogischen Praxis [The method of case reconstruction in basic research and in clinical and the pedagogical praxis]. In K. Kraimer (ed.) *Die Fallrekonstruktion. Sinnverstehen in der sozialwissenschaftlichen Forschung.* Frankfurt am Main: Suhrkamp.

Putnam, R. D. (1995) Bowling alone: America's declining social capital. *Journal of Democracy*, 6(1): 65–78.

Rammert, W. (2003) *Technik in Aktion: Verteiltes Handeln in soziotechnischen Konstellationen* [Technology in action: Distributed action in socio-technological constellations]. Berlin: Technical University Berlin, Technology Studies, Working Papers. Online, Available at: www.ts.tu-berlin.de/fileadmin/fg226/TUTS/TUTS_WP_2_2003. pdf (accessed 6 February 2017).

Reckwitz, A. (2005) Kulturelle Differenzen aus praxeologischer Perspektive: Kulturelle Globalisierung jenseits von Modernisierungstheorie und Kulturessentialismus [Cultural differences from a praxeological perspective: Cultural globalization beyond modernization theory and culture essentialism]. In I. Srubar (ed.) *Kulturen vergleichen. Sozial- und kulturwissenschaftliche Grundlagen und Kontroversen.* Wiesbaden, VS, Verlag für Sozialwissenschaften, pp. 92–111.

Sahakian, M. and Wilhite, H. (2014) Making practice theory practicable: Towards more sustainable forms of consumption. *Journal of Consumer Culture*, 14(1): 25–44.

Scammell, M. (2000) The internet and civic engagement: The age of the citizen-consumer. *Political Communication*, 17(4): 351–355.

Schneider, T. and Davis, T. (2010) Fostering a hunger for health: Food and the self in 'The Australian Women's Weekly'. *Health Sociology Review*, 19(3): 285–303.

Stolle, D. and Micheletti, M. (2003) The gender gap reversed: Political consumerism as a women-friendly form of civic and political engagement. Paper prepared for the Gender and Social Capital Conference, 2–3 May 2003. St. John's College University of Manitoba. Available at: http://umanitoba.ca/outreach/conferences/gender_socialcapital/ StolleMichelettipaper.pdf, (accessed 13 September 2013).

Stolle, D. and Micheletti, M. (eds) (2013) *Political Consumerism: Global Responsibility in Action.* New York: Cambridge University Press.

Stolle, D., Hooghe, M. and Micheletti, M. (2005 Politics in the supermarket: Political consumerism as a form of political participation. *International Political Science Review*, 26(3): 245–269.

Strübing, J. (2004) *Grounded Theory. Zur sozialtheoretischen und epistemologischen Fundierung des Verfahrens der empirisch begründeten Theoriebildung.* Wiesbaden: VS, Verlag für Sozialwissenschaften.

Swigger, N. (2013) The online citizen: Is social media changing citizens' beliefs about democratic values? *Political Behavior*, 35(3): 589–603.

Toffler, A. (1983) *Die dritte Welle, Zukunftschance* [The Third Wave]. München: Goldmann.

van Deth, J. (2014) A conceptual map of political participation. *Acta Politica*, 49(3): 349–367.

Waldschmidt, J., Klein, J., Korte, M. and Dalman-Eken, S. (eds) (2009) *Das Wissen der Leute. Bioethik, Alltag und Macht im Internet.* Wiesbaden: VS, Verlag für Sozialwissenschaften.

Witterhold, K. (2015) Politik des Lebensstils als eher weiblicher Partizipationsstil? Beteiligungspraktiken politischer Konsumentinnen und Konsumenten on/offline [Lifestyle politics as a rather female style of particiption? Participation practices of female political consumers on/offline]. *Forschungsjournal Soziale Bewegungen*, 2: 46–56.

Yang, M. and Baringhorst, S. (2014) Reintermediation durch Social-Web? Eine Analyse von Social-Web-Projekten im Bereich des politischen Konsums [Re-intermediation through the social web? An analysis of social web projects in the area of political consumption]. In F. Oehmer (ed.) *Politische Interessenvermittlung und Medien. Funktionen, Formen und Folgen medialer Kommunikation von Parteien, Verbänden und sozialen Bewegungen.* Wiesbaden: Nomos, pp. 399–423.

Yates, L. (2011) Critical consumption. *European Societies*, 13(2): 191–217.

6 Marketing critical consumption

Cultivating conscious consumers or nurturing an alternative food network on Facebook?[1]

Ryan Alison Foley

Introduction

Several anthropologists have recently documented the significance of local food movements across Italy, including Slow Food, Km0 markets, and alternative production models such as organic cooperative agriculture and 'co-production' (Black, 2012; Counihan, 2014a, 2014b; Grasseni, 2013, 2014a, 2014b; Leitch, 2003; Rakopoulos, 2014; Siniscalchi, 2012, 2014). Although the rise of political consumerism in Italy and elsewhere since the early 2000s has been attributed, at least in part, to the use of new information and communications technologies (ICTs) (Forno and Graziano, 2014), research on specifically *how* these technologies have been used to promote food activism in Italy is limited.[2] This chapter adds to this discussion by examining the social media marketing practices of an alternative retailer, Luminare cooperative, to consider how such retailer-based activism aims to promote alternative food networks by cultivating conscious consumers on Facebook. This case study reveals some of the limitations of this retail approach, while also underlining the need to consider social media usage within the context of wider communication strategies and business practices.

Luminare is a cooperatively owned retail centre in the city of Villacenseo,[3] near Bologna, that has as its mission the promotion of ethical or 'critical consumption' (*consumo critico*). The cooperative was founded in 2011 by 'a group of dreamers', as they referred to themselves, who were already active in the local cooperative movement as members of another cooperative called Verdecura. These dreamers wanted to create a new type of social enterprise, promoting more sustainable consumption by making various types of alternative products – organic, bulk, Fair Trade or Km0 – all readily available in one location. Unlike large food retailers that sell a wide range of goods, including organic, Fair Trade and budget products, or small specialist stores and local farmers markets with narrowly defined food agendas, Luminare promotes a multi-faceted vision of sustainable living through what they term a '360-degree' approach to informed purchasing. Although it is similar to some information-based approaches discussed below that aim to empower individual consumers, this form of retailer-based activism intends to provide consumers not only with information, but also with direct access to 'better' alternatives. It is also unlike more radical types of activism such as co-production that seek to disrupt consumer-producer relationships, in that the retail model continues to rely on the purchase decisions of individual consumers. Luminare's mission and details of the space, products, events and offers are displayed on their website, with Facebook posts used as constant reminders to help spread awareness. During the study, Luminare's management considered such forms of online engagement as free marketing tools that could help the cooperative to reach potential customers, although as will be discussed below, this strategy had limited impact. Rather than generating discussions and relationships with potential consumers online, the cooperative's posting and sharing practices appeared to enhance existing relationships and practices among an elite management group.

The following analysis is based on ten months of fieldwork from November 2013 to August 2014 during which I conducted participant-observation as a volunteer employee at Luminare and Verdecura cooperatives. I was uniquely placed to study Luminare's use of ICTs because I was tasked with weekly website updates for the Luminare events and Bio Grocer deals, and daily Facebook posts. Although I had direct access to the communication process and facilitated the social media marketing, I did not generate the content myself, but received text files from the events manager and the Bio Grocer manager detailing the upcoming events and deals, and occasional emails from the vice president with ad hoc content requests, such as sharing events and offers of other cooperatives in the local consortium. In addition to attending various meetings with the vice president and the communications director of Luminare, I regularly attended evening events at Luminare and was present daily in the shared administrative offices to observe and participate in casual discussions about the retail centre. These varied engagements allowed me to observe the cooperative's practices from multiple vantage points, both internal and external, online and offline.

Before going into greater detail about Luminare's communication strategy and how their ICT use developed, or in some cases failed to develop, relationships with consumers and producers in the context of this retail-based food activism, it is useful to review insights from previous studies on the intersection between ICT use and food activism. These examples shed light on the potentialities of ICT use in this context, and introduce the types of relationships that exist between diverse actors in the food system and some of the ways in which these have been challenged or mediated by ICT use.

Food activism, consumerism and ICT use in Italy

Food activism covers a wide range of practices, with each instance representing local circumstances and ethics. Carole Counihan and Valeria Siniscalchi, both anthropologists studying the Slow Food movement in Italy, have usefully defined 'food activism' as 'efforts by people to change the food system across the globe by modifying the way they produce, distribute, and/or consume food' (2014: 3). As such, food activism is often associated with 'alternative' practices, constructed in opposition to dominant – typically capitalistic – supply chains. In

Italy, for example, Slow Food grew out of a 1980s movement that opposed the opening of the first McDonald's fast food chain in Rome, and promoted local and small producers as an alternative to increasing homogenization (Ritzer, 2013: 148). Similarly, solidarity purchasing groups (*Gruppi di Acquisto Solidale*, GAS) promote 'short food chains' that support local as opposed to large producers (Grasseni, 2013: 5), and anti-mafia cooperatives aim to replace supply chains influenced by mafia-controlled agricultural industries with cooperatives that promote principles of solidarity and 'legality' (in essence, following both the letter and the spirit of the law) in addition to organic farming methods (Rakopoulos, 2014). Reflecting local concepts of right and wrong, and perceived problems in the existing food system, food activism thus covers a wide range of initiatives that take aim at various aspects of the food system from production to consumption processes. The goals of food activism range from fostering democracy, or more ethical and sustainable production, to healthier or higher quality foods (Counihan and Siniscalchi, 2014: 6), and include groups that span 'the entire [political] spectrum from left to right' (Wilk, 2006: 22).

Some strands of food activism follow a long tradition of ethical or political consumerism, which includes both boycotting and buycotting, reaching back to the nineteenth century or even earlier (Lewis and Potter, 2011), with the cooperative movement being sometimes considered as an early form of consumer advocacy (Gabriel and Lang, 2006: 9). There has, however, been a conflictual relationship between consumer cooperatives and consumer activism because while cooperatives have viewed their mutualistic model as the best way to protect the interests of the working class, consumer advocates have instead focused on lobbying, boycotting, and information-sharing geared to the individual consumer (Furlough and Strikwerda, 1999: 20). Some scholars have also questioned the efficacy of activism geared towards changing individual consumption behaviour. Writing about the genetically modified organism (GMO) movement, for example, Roff (2007) argues that consumerist tactics represent the neoliberalization of food activism, shifting responsibility from states and manufacturers to individuals. Labelling standards such as organic and Fair Trade also serve to enable individual consumers as agents of change, yet these alternatives to the standard food system have been, to varying extents, co-opted by the corporate food systems that such alternatives seek to challenge (Jaffee and Howard, 2009).[4]

It is thus important to distinguish ethical consumption practices, as a substrand of some forms of food activism, from ethical consumerism as an ideology (Littler, 2011). While some consumption is necessary for survival, consumerism represents a particular ideology, often linked to capitalism, as Robert Bocock (1993: 50) explains:

> that is the active ideology that meaning of life is to be found in buying things and pre-packaged experiences, pervades modern capitalism. This ideology of consumerism serves both to legitimate capitalism and to motivate people to become consumers in fantasy as well as reality.

Gabriel and Lang (2006) have identified five different strands of consumerist ideology, ranging from the dominant Western ideology linked to capitalism described above, to various 'alternative' ideologies, all of which grant power to the individual consumer to affect change. Alternative ideologies of ethical consumerism have grown since the 1980s and focus on the ability of the consumer to improve environmental, labour, health, or other ethical issues through advocacy and shopping choices (ibid.: 166). These, of course, are relevant concerns for many food activists, but the distinction is crucial because while some visions of an ethical food system may preclude consumerism as an ideological base of the dominant capitalist system, others, such as that promoted by Luminare, aim to use the power of the individual consumer to support alternative models of production. Indeed, a study of various types of sustainable community groups in Italy found that while a concern about the social and environmental impacts of materialism and consumerism is central, rather than being *anti*-consumption, the interest is in promoting artisanal products over mass-produced alternatives with a focus on fair distribution of wealth (Forno and Graziano, 2014: 143).

Other alternative practices directly target the distinction between producer and consumer and aim to reconfigure these relationships via alternative provisioning strategies such as co-production or cooperative production, as opposed to individual consumption. Cooperatives, as mentioned above, have promoted mutualism, aimed at changing patterns of ownership and blurring the boundaries between producer and consumer, since the late nineteenth century, thanks to an ideology of self-help (Gabriel and Lang, 2006: 157). Cristina Grasseni argues that co-production in solidarity purchasing groups (GAS) is also based on socio-economic mutualism (2013: 79). She conceives of their provisioning activism 'as a form of political activity, moving beyond the concept of consumer sovereignty' (ibid.: 78). Unlike seemingly similar community agriculture groups in the United States that enable direct links between consumers and local farmers without altering the underlying relationships between consumers and producers, co-production in GAS involves working directly with producers through regular meetings and voluntary work (Grasseni, 2014b: 179–180). However, the aims and ethics of various actors within the food system can be divergent, even within such groups ostensibly working together for change. Theodoros Rakopoulos' (2014) research on anti-mafia cooperatives emphasizes this point. While the cooperative administrators claimed to practise food activism and linked 'good' food with concepts like democracy, solidarity, legality and organic agriculture, the manual workers considered the wine produced by the cooperative to be 'too commercial' in opposition to their homemade wine that was 'authentic and pure' (ibid.: 117). Thus, Rakopoulos argues, even the claims of ethical production underscored existing divisions of labour that were perpetuated in the alternative model (ibid.). This highlights the difficulty of breaking away from established patterns of production when working within existing food systems.

In theory, using new information and communications technologies to support existing practices or new forms of food activism could help to establish novel communities of producers and consumers, though it is less clear how ICT use could challenge this basic distinction between them. Studies of ICTs and food in general have shown that ICT may enable new forms of sociality, particularly in the consumption sphere via access to information and the provision of information from diverse sources, including consumers themselves (Choi and Graham, 2014). However, research on such attempts to build new forms of sociality using ICTs problematizes their idealized nature. One example is *HowToBuyWiki*, which aims to promote conscious consumption by providing an open-source online platform where consumers are encouraged to compare various product attributes, including ingredients, size and organic certification. In this case, Eli, McLennan and Schneider (2015) argue that while the project of information sharing offers a form of cyber-care, the relationships that the members aim to develop with and among consumers are limited in practice by the anonymous use and rigid product attribute frameworks. Another example is the group of Buycott app users studied by Eli *et al.* (2016). The Buycott app collects crowd-sourced product information to support user-generated activist campaigns. However, despite claims to create a 'community', the commodity-centric construction of the ethical consumption app continues to promote an individualistic form of activism related to decision-making in retail stores (ibid.: 71–72).

In addition to various ways that ICTs may enable ethical consumption via access to or production of information, Berry and McEachern have also suggested that there is potential to 'create and reinforce "communities of interest"' (2005: 84). ICTs have the potential to alter spatial relations, either through the creation of virtual places or by connecting people across vast spaces and national boundaries (Green *et al.* 2005). Richard Wilk has suggested the importance of distribution strategies for connecting alternative consumers and producers, particularly in the face of the vast networks and sophisticated information technology used by the global food industry (2006: 22), yet he also warns that successful distribution strategies may reproduce capitalist relations between consumers and producers and commodification through marketing tactics, as seen in Lyon's (2006) ethnographic work on Fair Trade coffee. In the case of the anti-mafia wine cooperative studied by Rakopoulos, administrators used websites, e-newsletters and other tactics to build the reputation of the cooperative and to encourage distribution and consumption of their products, and this focus on the consumption end of the process, he argues, shaped the divergent food ethics between administrators and manual workers, thus reinforcing existing communities (2014: 115–116).

Other ethnographic cases, like the GAS network in Italy and the Luminare example below, show how ICTs can enable and support existing local networks. Grasseni's research on GAS suggests a minor role for ICTs in organizing their co-production practices. Although the network maintains a website and encourages member groups to register online, Grasseni highlights the relative importance of in-person meetings and the amount of offline work necessary to coordinate a collective purchase initiated by an online survey (2013: 99–100). Furthermore, she chronicles the rejection by some GAS participants of the use of online communication methods to coordinate large-scale orders, as these participants argued that online communication was antithetical to their ideal of supporting small local producers (ibid.: 105).

Social media marketing, which figures centrally in the Luminare case, also reflects local practices and concepts, despite the potential of global reach. The concept of marketing itself is culturally contingent, and intended to produce markets for goods or services (Araujo *et al.*, 2010: 1). Social media marketing is a new form of marketing that, in opposition to the traditional communication methods of mass marketing, has the potential to reach consumers on a more individualized basis via social media, such as blogs, forums and social networks (Evans, 2010). Miller *et al.* found engagement with advertising on Facebook to be functional primarily at the local level, mostly useful for 'commercial enterprises that are themselves based on personal connections and small-scale sociality' (2016: 97). In southern Italy, social media use was common among entrepreneurs and part of a wider context of cultivating relationships and personal visibility (ibid.: 93). Indeed, as will be seen below, Luminare's use of Facebook marketing showed a greater potentiality for maintaining existing network ties than it appeared to have for cultivating new consumers.

In order to understand the impact of ICT use on Luminare's practice of food activism, I first consider what makes the Luminare retailer model innovative, according to is founders. Next, I discuss how Facebook marketing fits into Luminare's broader communication strategy, before looking at how the alternative foods on offer at Luminare were advertised through Facebook posts. Finally, I describe the complete communication process from Facebook through to an in-person event, to indicate how the cooperative leaders aimed to modify relations within the food system, an aspect that may not be entirely clear if considering only the online practices.

Luminare's 360 degrees of critical consumption

The Villacenseo city centre is small enough to walk across from one side to the other in less than an hour. The central *piazza* is surrounded by old *palazzi*, a post office and various other public buildings of light grey stone embellished with multi-storey columns and grand arched windows. The route from the train station to the offices of Luminare each morning would take me past office buildings of modern brick and converted factories, many small shops, and multiple grocery stores: a large Coop supermarket, part of the Coop Italia network of consumer cooperatives; an A&O supermarket chain store; and the discounter Lidl. Dotted throughout the city were other small discounters and supermarket chains, CONAD retailer cooperatives, an EATaly chain specializing in Italian products, and three small organic stores. Aside from the specialty stores, Coop, CONAD and the other large retail chains like A&O also sold organic and Italian specialty ranges. Even the German discounter Lidl actively promoted the Italian provenance of many of their goods. Until 1993, organic products were available only in specialty stores; availability was extended to national supermarket chains first by Coop Italia, and it was estimated that 5 per cent of total Italian food consumption would be organic by 2010 (FAO, n.d.). Despite the increase in availability, retail prices for organic fruits and vegetables ranged from 50 to 200 per cent higher than standard (ibid.), and thus may still be out of reach for many consumers.

Thrift remains a key element for many Italian shoppers, just as Daniel Miller observed in his study of shopping in north London (1998). A clear distinction exists in Italy between shopping for necessities (*fare la spesa*), or grocery shopping, and leisurely shopping for other items (*fare lo shopping*). On Sundays, many people in Italy *fare lo shopping* with friends or family, and this is a distinct activity from *fare la spesa*, which is linked to the idea of necessary expenses, *spesa*, being also used to denote bills such as heat and electricity. This difference appeared to coincide with parts of the Luminare business that were flourishing (consumables linked to daily necessity), versus those that were struggling (clothing and other more durable items). A recent study of market shoppers in Turin found that while grocery shopping, there is a focus on thrift and justifications are given for occasional sweet treats, shopping at the more expensive Sunday market which features organic and specialty items is considered in itself to be 'a treat, a leisure activity, rather than a necessity' (Black, 2012: 71, 83). Although awareness of food origin and organic production has been spreading in Italy because of the visibility of the Slow Food movement and also encouraged by food labelling laws introduced in 2002, price is still seen as the bottom line for many shoppers (ibid.: 148–149). Indeed, the European Social Survey found that in cases where ethical products cost more than standard, their purchase is correlated with middle-class social status (Harrison, 2010).

Consumer and retail cooperatives are prevalent in Italy, representing about one-third of the total grocery market share in 2015.[5] Coop Italia invests heavily in education and outreach to promote conscious consumerism and their logo can be frequently seen on banners and programmes supporting local events, ranging from concerts and book fairs to the annual 'Week of good living' held in nearby Forlì. As mentioned above, they were the first supermarket chain in Italy to begin carrying organic produce. They had more recently introduced two additional tiers of own-label products: *vivi verde* (an organic line) and *fior fiore* (a gourmet line focusing on local specialty items) to enhance their reputation for safety and quality, according to a board member of one of the local Coop branches near Bologna. Coop Italia also promotes various types of ethical consumption in store, including healthy eating for children.

The act of prominently displaying the Coop stance regarding local food, animal testing, genetically modified organisms and responsible fishing both adds to the Coop brand image and provides information to consumers who are concerned about these issues (Figure 6.1). Likewise, the Coop offers 'Super Vitamini' plush toys and books that simultaneously promote awareness of healthy foods for kids, while feeding into a customer loyalty programme whereby the plush toys are bought with a combination of money and qualifying purchases, or else entirely using membership points.

Figure 6.1 Signs in a Coop store. Top row (left to right): 'Responsible fishing'; 'No thanks to animal testing'; 'No thanks to GMOs' and '100% Italian olive oil'.

Source: author's own photograph.

Unlike Coop Italia, a consumer cooperative, Luminare is a social cooperative, a non-profit business designed to provide work placements and jobs for disadvantaged members of society, including disabled people and people with long-term unemployment. Social cooperatives emerged in Italy in the 1980s, filling gaps left by the retreating welfare state to provide care or work integration for individuals who had previously been in care.[6] Catherine Trundle (2014) has read this as part of an increasingly neoliberal model of government decentralization. However, to early adopters of the social cooperative form, this shift also represented the culmination of a liberal movement that sought greater equality and social inclusion (Marzocchi, 2014) and supporters of the cooperative model have seen it as an innovative solution to ensure care provision through civil society (Restakis, 2010). Regardless of whether social cooperatives are seen as part of a neoliberal programme of disintegration at the state level, or as an alternative model of integration at the social level, by including people in the work force who previously relied on family or public support, these cooperatives enable more people to earn a regular income as workers, and spend it as consumers. Cooperatives like Luminare are likewise simultaneously catering to the needs of members through the provision of work, while striving to answer the needs of consumers through goods or services sold on the market. This is a new element for some social cooperatives that had previously relied heavily on grants and public contracts, which are becoming less readily available, thus requiring them to compete with other businesses for a share of consumer spending.

The founders of Luminare saw the venture as a means through which they could spread their own values of critical consumption by providing a physical space where consumers would have immediate access to a range of informed purchasing options. The mission of Luminare cooperative, which serves as the guiding principle for the retail space, is shared in the 'About Us' section of their website:[7]

> Luminare is a project developed for the local community, dedicated to the social economy and the promotion of human and economic activities, *sustainable* from an *ethical, social* and *environmental* point of view, and which do not have profit as the main priority and reason for being. All activities must have as their primary goal the enhancement of the capacity of each person, to be directed to the affirmation of the principles of *solidarity* and *social justice* and pursue the *greater good* for the greatest possible number of people. Any initiatives that do not show *respect* for *human rights, animals*, and *nature* are also prohibited.

> (Emphasis added)

Consistent with other examples of ethical consumption, this vision includes a diverse range of overlapping and potentially conflicting concerns: economic, environmental, social, and animal, as well as potentially conflicting views on territoriality and a concern for solidarity and the greater good.

Luminare itself is located about 5 km outside the city centre, closer to the farms and orchards that are the engine of the Emilia-Romagna countryside than to the other shops and markets in the centre of the city. At the time of the study, there was ample parking space in front of the building and a small bike rack, which more often than not remained unused. The nearest bus route was rather inconvenient. Most customers and the cooperative members working there arrived by car. This posed a problem for the management in terms of attracting custom from the city centre, so much of the physical advertising of printed leaflets and newsletters was geared towards the surrounding neighbourhoods. However, the environmental impact of the reliance on individual transport did not appear to be a concern, despite the obvious conflict with their promotion of Km0 foods and the spring water dispenser that eliminated transportation of pre-bottled water.

Much of the open plan space at Luminare was decorated with reclaimed wood, 'upcycled' items like lights made from wine bottles, and vintage or artisanal furniture and artwork. This created a certain aesthetic, which the vice president Giada explained to me, was an alternative look, yet one that would hopefully not alienate fashionable women. Women in Italy, as elsewhere, have traditionally been responsible for the unpaid work of social reproduction, including shopping and food preparation (Thelen, 2015), and have thus been at the forefront of many grassroots practices of food activism (Counihan, 2014b). It is also an arena that, according to Counihan, gives women an opportunity to express their agency both publicly and privately, though she calls into question the extent to which this can change society at large, including the gendered division of labour itself (ibid.: 63, 99). Giada envisioned Luminare as more than 'just about being "good" but [also] showing that it is possible to be innovative, fresh, and responsible with higher quality products and with glamour'. Events such as vintage-inspired fashion shows, featuring sustainable, upcycled or reused garments, organic cosmetic beauty evenings, and cooking classes for children were also part of this outreach to women and families.

What makes Luminare innovative, according to the management, is the 360-degree approach that combines various types of critical consumption under one roof. Shortly after opening in 2012, the communications director noted the importance of the shopping centre format in their vision: 'Instead of proposing yet another small organic store, we asked ourselves why it wouldn't be possible to take advantage of the strengths of a shopping centre – concentrating multiple types of products [in one space] – but to transform it ethically.' The shopping mall is a way to increase 'efficiency' for both retailer and consumer by attracting crowds of people and providing numerous shops and activities in one place (Ritzer, 2013: 59). Beyond the added convenience, George Ritzer theorizes that the shopping centre is an important symbol in a culture of consumerism (2005: 7). In his analysis of such spaces, Ritzer highlights how the structure and ambiance of retailers from different eras in Milan influence how consumers practise and experience consumption (ibid.: 156–159). The open-plan space with the Km0 bistro and Bio Grocer at the back encouraged Luminare's visitors to browse effortlessly from the vintage corner to the organic cosmetics on their way to buy instant consumables. This desire to create synergy, and encourage shopping, is also reflected in their communication strategy below, through the use of events and offers to drive footfall and potential sales. While espousing critical *consumption* in their mission, the physical space and online marketing practices seem to reflect the ideology of *consumerism* where customers are encouraged to participate through shopping.

An integrated communication strategy

The website and Facebook page were central to Luminare's communication strategy because they could, in theory, generate awareness of their products at virtually no cost, one of the potentialities of ICT noted by Berry and McEachern (2005). ICTs were part of a broader integrated strategy that included various traditional and new media. Online channels, notably Facebook and the Luminare website, were used to communicate the planned events and the weekly offers that were also published in a paper and e-newsletter, and occasional press releases (frequently also online). Luminare's management conceived of their Facebook presence primarily as a marketing vehicle. Although the stated mission of the cooperative was to promote 'human and economic activities, sustainable from an ethical, social and environmental point of view, and which do not have profit as the main priority and reason for being' (as described in their mission statement above), Facebook activity was intended to cultivate critical consumers, with an emphasis on *consumers*, to drive Luminare's sales and make the retailer itself profitable, or as more regularly termed, financially sustainable.

Facebook marketing: 'Using the tools of the big guys'

One afternoon I was waiting at Luminare for a communication meeting with the web designer Nico, Luminare's vice president Giada, event planner Lucia and volunteer Rachele, who was also going to be helping with the communication efforts. Rachele and I sat together and discussed the marketing strategy and questions for Nico, who would be training us later that afternoon. We would soon be taking over the external communications: she would be working on the bi-monthly newsletter and press releases, and I would be doing the weekly website updates and Facebook posts. We also considered how we might make the communication more successful and attract more customers. 'We need to use the same tools as the large corporations, but in our own way,' Rachele suggested. Verdecura's director Mattia told me previously that it would be necessary to '*fare il ruffiano*' (be sly or pandering), taking a marketing-oriented approach when posting on Facebook. As an example, he jokingly suggested that we could post kittens to catch attention even if not relevant;[8] the social media content was seen as less important than the outcome: generating page likes and thus casting a wider marketing net. One day, a few months later, I congratulated Lucia and her assistant as they practically danced around the shop floor after the Luminare page likes reached 1,000.

When we eventually sat down with Giada, she stressed the importance of the added value of the goods they sold. The communication strategy would have to address a general impression that the Bio Grocer was an expensive place to buy food. The approach would be two-sided: 'Why does it cost more?' and 'Does it *really* cost more?' How could we communicate this? I suggested a direct comparison to other stores. Recalling Mattia's comment on my first visit to Luminare that some of the basic organic goods were a similar price to regular branded products like Barilla pasta, I suggested that we could provide a comparison of a similar basket of products to show that the absolute price was not very different, as a starting point to combat price-value concerns. Giada thought that this would be too direct and aggressive. Instead, she suggested using the website and e-newsletter to focus on quality, and on the added ethical value, created through relationships with producers. Facebook could be used to create more publicity, to really 'bombard' people with messages as it was free. She outlined the recommended practice for the posts. For each event, there should be three reminder posts: two days before, the day before and the same day. The length should be 60–70 characters maximum, and she suggested using the hashtag symbol and capitalization to emphasize important words. Each post should have an image to grab attention or a link pointing to the website to drive traffic. Furthermore, I could also share posts from other relevant pages, for example, Greenpeace. The test for posting was, 'Would I be interested to read this?' Nico added, 'If the page is "us, us, us, us" all the time, it doesn't work very well', so varying the content in that way would help to keep people interested. This mirrored the suggestion from *Social Media Marketing for Dummies* 'to share in the spirit of giving' in order to encourage other users to share your posts (Khare, 2012: 10), and displayed Nico's specialist knowledge of social media marketing theory and praxis.

The resulting interactions via social media reflect Luminare's marketing approach, as Facebook was mostly used to promote events and share grocery offers, other new products or special deals, all tactics to drive footfall. The weekly Km0 market reminder post is typical, consisting of an image of the market or of fruits and vegetables – it was considered important to regularly change the image to catch attention – and highlighting the local nature of goods through the use of the Km0 label:

TODAY 14:30–18:30 Km0 organic market with local producers
Fruit, vegetables and many other specialties directly to your table!

Likewise, the separate Bio Grocer page shared weekly specials, recipe ideas and often reposted the Luminare events. Below are some typical posts of products on offer. These also relied on labels, in this case, 'organic', to promote the ethical added value, yet in order to drive sales the primary focus of the posts is on price reduction. In both cases, the discount is written in capital letters for emphasis, while only in the second example is the organic label also capitalized and given equal importance.

THE SUPER OFFER of the Luminare Bio Grocer for 10–15 JUNE!
Organic Prosciutto Crudo – EXCEPTIONAL 25% DISCOUNT: only €2.15 per 100g!
To see all the offers > web link
The new SUPER OFFER has arrived in the Luminare Bio Grocer!
ORGANIC sparkling white wine from the Azienda Agricola Fiammetta – PAY ONLY ONE EURO FOR THE THIRD BOTTLE WHEN YOU BUY TWO!!
> From the clear notes of the pignoletto and trebbiano [grapes] are born the pleasing bubbles of Clarinetto! A wine to drink in the country, ideal as an aperitif, with fish, or the whole meal.
OFFER VALID from 10 to 24 May as supplies last.

Such posts did not generate discussions or debate online, but represented a one-way flow of digital information. The Bio Grocer page had around 400 followers and over the ten months that I observed there was only one post to the page from a would-be consumer. Other posts were from other cooperatives in the network, extending their reach among the followers of each other's pages. In a typical example of a Luminare Facebook post, there were four post shares, all of them by other local cooperatives in the same consortium.

Giada, Lucia or Mattia would often like the Facebook posts, and I wondered if it was their way of recognizing my efforts, or rather an attempt to boost their visibility via assumptions of how the Facebook algorithms might work.[9] Sharing posts was also frequently done by certain individuals, but I recognized them as those who worked in communications or other management positions within the cooperative or broader consortium. This could be viewed positively as cooperative behaviour intended to help promote the messages of their peers, or, from a more cynical viewpoint, could be read as self-interested behaviour to strengthen their own network connections. In either case, the writing and sharing of posts appeared to represent a top-down flow of marketing information from an elite group of administrators, as opposed to creating a dialogue towards a shared view of sustainable living.

Facebook within the broader communication strategy

However, Facebook was only one part of a larger communication strategy that included the Luminare website, (e-)newsletter, weekly events and press releases such as the one mentioned above. The newsletter draft was the central source of inspiration for online content, containing all announcements of events and discounts and short articles on critical consumption. Each newsletter had a similar format of four colourful pages (two external and two internal), also published virtually on the website and distributed via email. The front page was filled each time with a large image corresponding to a short editorial piece, highlighting a particular product or service at Luminare. A selection of front-page editorial topics included: vintage and restoration; eco-fountain filtered water distribution; oranges for health and legality; clothing made from recycled fabrics. The back page comprised a selection of vouchers and announcement of current offers in the Bio Grocer. Inside was a calendar of events, a recipe suggestion (using ingredients available in the Bio Grocer), and a few short feature articles, also promoting products or services available. Examples include: 'Some recommendations for a GREEN WARDROBE'; 'We are what we eat'; 'Why organic undergarments?'; 'Why buy bulk goods?'; and various cosmetics articles related to the season such as 'Summer skin protection' and 'SOS for cold weather'. Although these articles promoted various aspects of Luminare's vision of ethical shopping, they are all instances of marketing, encouraging some sort of consumption rather than including other tips for sustainable living such as reduction, re-use or self-sufficient production. The articles, again, appeared to promote ethical *consumerism*, rather than a version of critical consumption that included various acts of non-consumption or questioned the role of the consumer.

Nevertheless, looking at one example to see how it was communicated on Facebook, in the newsletter, and in person at the event itself, shows that taking into consideration *only* the Facebook posts outside the context of Luminare's broader communication strategy could exaggerate the consumerist aspect of the initiative. In early 2014 the cooperative held a small convention entitled 'The Fruits of Legality' with four speakers: a lecturer on social enterprise and three presidents of social cooperatives, including a citrus cooperative that participates in the anti-mafia cooperative agriculture movement. The Facebook post, limited in length, provided only the date, title and list of speakers. Thus, an analysis of the Facebook posts alone would be unlikely to provide a useful understanding of the concept of fruits of legality. The newsletter, on the other hand, featured an article entitled 'Health, quality and legality, all in an ORANGE' that fleshed out the concept in more detail. In addition to promoting the taste and health benefits of the organic oranges – sold exclusively in the Luminare Bio Grocer, the article noted – it provided some explanation of the legality aspect of the social cooperative:

> These oranges 'are good' (*fanno bene*) not only for the people who eat them, but they are also the fruits of the lands of legality. From Locri to the plains of Gioia Tauro extend hectares of land confiscated from the mafia and reha-bilitated by producers who have chosen to join the GOEL consortium.

This begins to paint a picture of the alternative production model implied by 'The Fruits of Legality' label.

For the twenty-odd people who actually attended the event, the meaning and implication of the concept of legality were again further elaborated. The president of the citrus cooperative focused on creating a model that was not only more ethical in its business practices, but also waas seen to be more effective than the previous one, with mafia involvement perverting the 'proper' functioning of the market. Whereas previously producers could earn only about 5 cents per kilo of fruit, the minimum price to producers in the consortium was 40 cents. Those who stood up against the mafia as part of the consortium risked the destruction of their property though fires, but they could also expect to earn eight times more. Rakopoulos has written about legality cooperative members being convinced about 'using only the wallet' as opposed to indoctrinated into a shared vision food ethics (2014: 117), but in the view of this citrus cooperative president, like the administrators in Rakopou-los' case, this mode of production was a tool for cultural and political change, which required offering a viable alternative. His hope was that as more people began to move towards spending their own money ethically, it would transform society. Luminare, he said, was going in the direction that the whole economy needs to – and will, he hoped – move towards. The aim was to empower people, as consumers, to contribute to change through their shopping choices, and while GOEL provided the alternative production model, retailers like Luminare could provide the link between producers and consumers. Therefore, cultivating critical consumers was necessary to sustain both Luminare's own alternative retail model and to support those of the alternative producers that they stocked. While not chal-lenging the fundamental roles of consumer, producer and retailer, the hope was to orient them all towards more ethical and sustainable practices as detailed in Lumin-are's mission statement.

Discussion: Luminare's consumerist vision of food activism

Although in some areas Luminare did encourage using less (via reduced pack-aging and upcycling), the main thrust of their communication remained consum-erist, with a focus on discounts and product features, and not all consumers could choose to pay the higher prices for organic produce, limiting the efficacy of this approach to change. While there are some who can choose to spend 'more ethi-cally', for others, it is not only a question of ethics, but of basic affordability. To illustrate this, consider the differences between a typical Luminare shopper and some of the cooperatives' employees. I became friendly with a local entre-preneur who lived just outside Villacenseo, not far from Luminare. She and her family were part of the core target of the Luminare Bio Grocer, that is, local families. She and her husband shopped there regularly, buying basic groceries, paper products and cosmetics. Their grocery shopping repertoire also included organic goods from CONAD and Coop and produce from local markets. They shopped at Luminare without appearing to question the price; I watched as they dropped various goods into their basket without checking price tags. Alterna-tively, most people working in Verdecura and Luminare cooperatives felt they could not afford to pay more for many of the goods being sold there. I heard fre-quent laments that the goods were nice, but just too expensive. One of the office workers, Serena, could justify paying a higher price for organic olive oil that she considered tastier and healthier and could be used at every meal, but she could not make the same argument for other goods, as an excerpt from my field notes, capturing an exchange between Serena and Lisa in late 2013, shows:

> Serena is saying she doesn't shop at Luminare because it really costs too much. She describes something she wants to buy and the price, but, 'No'.
>
> She can buy something in a small store nearby for much less. 'I know that many people go there and spend a lot, but I feel like saying, "Eh ..."'
>
> Lisa: 'I'm not sure what kind of clients can shop there.'
>
> Serena: 'People with lots of money who don't care about the cost.'
>
> Both agree that even if they agree in principle, they can't afford to shop there.

Lucia, the Luminare events planner, was also tasked with weekly deliveries of organic groceries to the offices in the consortium, though her Friday shopping deliveries only went to people like Giada and Mattia – those at the highest man-agement levels. Although, when I first met him, Mattia had pointed out to me the small disparity at Verdecura between the lowest-paid and the highest-paid employees, about 35 per cent difference in his estimation, this discrepancy of €300–400 per month was enough to make a difference in discretionary spending – the difference between buying only on the basis of price, and having room in the budget to choose to pay extra for added value(s). In the political consump-tion repertoire of many of my colleagues, boycotting and other forms of reduced consumption were readily available, whereas making the more ethical purchase often was not.[10] While the social cooperative was enhancing consumer purchasing power via work placement salaries, ethical purchasing options were not equally available, and Luminare was marketing added value to a small group of affluent consumers. Making the alternative products available and increasing awareness of them via social media marketing did not address the root problem of some consumers.

Despite the apparent goal of empowering consumers, Luminare's Facebook marketing revealed a top-down flow of information, with a small group of individuals promoting their own visions of ethical consumption, and a strategy that focused on driving sales to maintain their own retail-based approach. As detailed above, Facebook was viewed primarily as a marketing tool to attract customers to the retail space with attention paid to the total number of page likes, and content being sometimes less important than gaining visibility. With the primary focus being on deals and events to drive footfall, Luminare appeared to promote ethical consumerism as opposed to other types of food activist practices that directly challenge consumer/producer relations. Furthermore, despite discussions of added value, the weekly promotions shared on Facebook regularly used price discounts, again showing the limitations of relying on individual purchasing power to fundamentally change the food system. The patterns of reciprocal sharing and liking on Facebook showed that the relationships created and maintained were primarily among a network of local cooperatives, and more specifically among their management. They reflected established bonds resulting from existing contact, and thus mirrored the aspirations and forms of activism available to an elite group of managers and not even all members of the two cooperatives. While Luminare's use of Facebook conformed in some ways to the platform's dominant usage modes, for example, favouring images, sharing of other links to get attention, and using short commentary with #hashtag on keywords like #bio or #km0, it also did not take full advantage of some of the potentialities of the platform, such as its international reach, and Facebook was used primarily to spread awareness and facilitate local action, a pattern also observed elsewhere (Grasseni, 2013; Valluari, 2014; Miller *et al.*, 2016). Confirming Miller *et al.*'s (2016) findings, Luminare's use of social media appeared to strengthen existing bonds, nurturing a network primarily among other local cooperatives, rather than cultivating consumers, which ostensibly was at the core of the project.

Despite the limitations of the approach, Luminare's use of Facebook marketing must also be understood in the context of their broader communication strategy and mission. Using Facebook to promote their goods through discounting may appear to de-value them, and relying on labels like Fair Trade and Km0 may ultimately reproduce processes of commodification of the products they sell (Freidberg, 2003), but Luminare's approach takes advantage of these existing paradigms to maintain their own alternative distribution model, and consequently to support alternative producers. Luminare's sales were contributing to creating jobs locally for disadvantaged workers and contributing to the sales of other local producers and artisans. While only a partial solution and one that furthers divisions between those who are able to choose to purchase more ethical goods and those who are not, promoting such alternative production and retailer models may help to create a sense of diffuse solidarity (Forno and Graziano, 2014), and represent the beginning rather than the end of processes of change.

Acknowledgements

I am very grateful to the editors of this volume for their feedback on early drafts of this chapter. My PhD supervisor Robert Parkin and mentor Inge Daniels have also provided valuable input to help think through these issues, and of course I cannot thank enough the members, staff and shoppers of Luminare cooperative, who welcomed me into their workplace and homes.

Notes

1 This chapter is based on a PhD thesis chapter entitled 'Added-value: The limits of consumer democracy', submitted in April 2017.
2 Grasseni (2013) and Rakopoulos (2014) do make some mention of ICT use, which I will discuss below.
3 Both the name of the cooperative and the name of the city are pseudonyms. Similarly, names of individuals used throughout this chapter have been changed.
4 Research on the use of ethical food labels in Great Britain also indicates that these labels are themselves fetishized, having become tradeable commodities in their own right (Freidberg, 2003).
5 See www.alleanzacooperative.it/l-associazione (accessed 25 April 2017).
6 The two types of social cooperatives are labelled as Type A (care provision) and Type B (work integration).
7 I have translated this text from Italian for the purposes of this chapter. The same is true for subsequent quotations from personal discussions, Facebook posts and other online content.
8 This phenomenon, known as 'cat content' is a recognized way to garner attention on social media and other online platforms (Podhovnik, 2016).
9 In the Luminare and Verdecura offices we passed around the link below to a blog post describing the Facebook algorithm changes: www.slate.com/blogs/business_insider/ 2014/02/14/how_the_new_facebook_news_feed_works.html (accessed 20 February 2017). While ostensibly focused on the quality of posts, we noted that it coincided with prospecting calls from the Facebook marketing team.
10 This issue also extended to those working in other cooperatives, for example, Chiara, who worked at a national worker cooperative based in Bologna. She had worked there for about two years, working through three temporary contracts before being offered a permanent one, yet still she was unable to afford any savings after her monthly expenses were paid. 'At the end of the month there is nothing left,' she told me.

References

Araujo, L., Kjellberg, H. and Finch, J. (2010) *Reconnecting Marketing to Markets*. Oxford: Oxford University Press.
Berry, H. and McEachern M. (2005) Informing ethical consumers. In R. Harrison, T. Newholm and D. Shaw (eds) *The Ethical Consumer*. London: Sage, pp. 69–87.

Black, R. (2012) *Porta Palazzo: The Anthropology of an Italian Market.* Philadelphia, PA: University of Pennsylvania Press.

Bocock, R. (1993) *Key Ideas: Consumption.* London: Routledge.

Choi, J. H-J. and Graham, M. (2014) Urban food futures: ICTs and opportunities. *Futures*, 62(B): 151–154. Available at: https://ssrn.com/abstract=2515384 (accessed 20 February 2017).

Counihan, C. (2014a) Cultural heritage in food activism: local and global tensions. In M. Di Giovine and R. Brulotte (eds) *Edible Identities: Food as Cultural Heritage.* Farnham: Ashgate, pp. 219–230.

Counihan, C. (2014b) Women, gender, and agency in Italian food activism. In C. Counihan and V. Siniscalchi (eds) *Food Activism: Agency, Democracy and Economy.* London: Bloomsbury, pp. 61–76.

Counihan, C. and Siniscalchi, V. (2014) Ethnography of food activism. In C. Counihan and V. Siniscalchi (eds) *Food Activism: Agency, Democracy and Economy.* London: Bloomsbury, pp. 3–12.

Eli, K., Dolan, C., Schneider, T. and Ulijaszek, S. (2016) Mobile activism, material imaginings, and the ethics of the edible: Framing political engagement through the Buycott app. *Geoforum*, 74: 63–73. Available at: www.sciencedirect.com/science/article/pii/S0016718515301251 (accessed 20 February 2017).

Eli, K., McLennan, A. and Schneider, T. (2015) Configuring relations of care in an online consumer protection organization. In E. J. Abbots, A. Lavis and L. Attala (eds) *Careful Eating: Bodies, Food and Care.* Farnham: Ashgate, pp. 173–194.

Evans, D. (2010) *Social Media Marketing: An Hour a Day.* Hoboken, NJ: John Wiley & Sons, Inc.

FAO (Food and Agriculture Organization) (n.d.) World markets for organic fruits and vegetables: Italy. Economic and Social Development Department. Available at: www.fao.org/docrep/004/y1669e/y1669e0a.htm (accessed 20 February 2017).

Forno, F. and Graziano, P. (2014) Sustainable community movement organisations. *Journal of Consumer Culture*, 14(2): 139–157. Available at: http://journals.sagepub.com/doi/full/10.1177/1469540514526225 (accessed 20 February 2017).

Freidberg, S. (2003) Cleaning up down south: Supermarkets, ethical trade, and African horticulture. *Social and Cultural Geography*, 4(1): 27–43. Available at: www.tandfonline.com/doi/abs/10.1080/1464936032000049298 (accessed 20 February 2017).

Furlough, E. and Strikwerda C. (1999) Economics, consumer culture and gender: an introduction to the politics of consumer cooperation. In E. Furlough and C. Strikwerda (eds) *Consumers Against Capitalism? Consumer Cooperation in Europe, North America, and Japan, 1840–1990.* Boston: Rowman & Littlefield, pp. 1–20.

Gabriel, Y. and Lang, T. (2006) *The Unmanageable Consumer*, 2nd edn. London: Sage.

Grasseni, C. (2013) *Beyond Alternative Food Networks: Italy's Solidarity Purchase Groups.* London: Bloomsbury.

Grasseni, C. (2014a) Food activism in Italy as an anthropology of direct democracy. *Anthropological Journal of European Cultures*, 23(1): 77–98. Available at: www.berghahnjournals.com/abstract/journals/ajec/23/1/ajec230105.xml (accessed 20 February 2017).

Grasseni, C. (2014b) Seeds of trust. Italy's *Gruppi di Acquisto Solidale* (Solidarity Purchase Groups). *Journal of Political Ecology*, 21: 178–192. Available at: http://jpe.library.arizona.edu/volume_21/Grasseni.pdf (accessed 20 February 2017).

Green, S., Harvey, P. and Knox, H. (2005) Scales of place and networks: an ethnography of the imperative to connect through information and communications technologies.

Current Anthropology, 46(5): 805–826. Available at: www.journals.uchicago.edu/doi/full/10.1086/432649 (accessed 20 February 2017).

Harrison, R. (2010) Ethical consumption and social class, *Ethical Consumer*. Available at: www.ethicalconsumer.org/commentanalysis/consumerism/ethicalconsumptionamiddle-classdistraction.aspx (accessed 20 February 2017).

Jaffee, D. and Howard, P. H. (2009) Corporate cooptation of organic and fair trade standards. *Agriculture and Human Values*, 27(4): 387–399. Available at: www.fairtradewire.com/wp-content/uploads/2010/09/Jaffee-Howard-2010-Cooptation-AHV1.pdf (accessed 20 February 2017).

Khare, P. (2012) *Social Media Marketing for Dummies*. Hoboken, NJ: For Dummies.

Leitch, A. (2003) Slow food and the politics of pork fat: Italian food and European identity, *Ethnos*, 68(4): 437–462.

Lewis, T. and Potter, E. (2011) Introducing ethical consumption. In T. Lewis and E. Potter (eds) *Ethical Consumption: A Critical Introduction*. London: Routledge, pp. 3–24.

Littler, J. (2011) What's wrong with ethical consumption? In T. Lewis and E. Potter (eds) *Ethical Consumption: A Critical Introduction*. London: Routledge, pp. 27–39.

Lyon, S. (2006) Just Java: Roasting fair trade coffee. In R. Wilk (ed.) *Fast Food/Slow Food: The Cultural Economy of the Global Food System*. Plymouth, UK: Altamira Press, pp. 241- 258.

Marzocchi, F. (2014) A brief history of social cooperation in Italy: looking to the future. *Aiccon* [ebook]. Available at: www.aiccon.it/en/pubblicazione/ (accessed 20 February 2017).

Miller, D. (1998) *A Theory of Shopping*. Cambridge: Polity Press.

Miller, D., Costa, E., Haynes, N., McDonald, T., Nicolescu, R., Sinanan, J., Spyer, J., Venkatraman, S. and Wang, X. (2016) *How the World Changed Social Media*. [ebook] London: UCL Press. Available at: www.ucl.ac.uk/ucl-press/browse-books/how-world-changed-social-media (accessed 20 February 2017).

Podhovnik, E. (2016) The meow factor: An investigation of cat content in today's media. In *Proceedings of Arts & Humanities Conference*, Venice, pp. 123–135. Available at: www.iises.net/proceedings/arts-humanities-conference-venice/table-of-content/detail?article=the-meow-factor-an-investigation-of-cat-content-in-today-s-media-6257 (accessed 25 April 2017).

Rakopoulos, T. (2014) Food activism and antimafia cooperatives in contemporary Sicily. In C. Counihan and V. Siniscalchi (eds) *Food Activism: Agency, Democracy and Economy*. London: Bloomsbury, pp. 113–124.

Restakis, J. (2010) *Humanizing the Economy: Co-Operatives in the Age of Capital*. New York: New Society Publishers.

Ritzer, G. (2005) *Revolutionizing the Means of Consumption: Enchanting a Disenchanted World*, 2nd edn. London: Pine Forge Press.

Ritzer, G. (2013) *The McDonaldization of Society*, 20th anniversary edn. Los Angeles: Sage.

Roff, R. J. (2007) Shopping for change? Neoliberalizing activism and the limits to eating non-GMO. *Agriculture and Human Values*, 24(4): 511–522. Available at: https://link.springer.com/article/10.1007/s10460-007-9083-z (accessed 20 February 2017).

Siniscalchi, V. (2012) Al di là dell'opposizione slow-fast. L'economia morale di un movimento. *Lo Squaderno*, 26: 67–75. Available at: www.losquaderno.professionaldreamers.net/wp-content/uploads/2012/11/losquaderno26.pdf (accessed 20 February 2017).

Siniscalchi, V. (2014) Slow food activism between politics and economy. In C. Counihan and V. Siniscalchi (eds) *Food Activism: Agency, Democracy and Economy*. London: Bloomsbury, pp. 225–241.

Thelen, T. (2015) Care as social organization: Creating, maintaining and dissolving significant relations. *Anthropological Theory*, 5(4): 497–515. Available at: http://journals.sagepub.com/doi/abs/10.1177/1463499615600893 (accessed 20 February 2017).

Trundle, C. (2014) *Americans in Tuscany: Charity, Compassion and Belonging*. Oxford: Berghahn Books.

Valluari, U. (2014) Transition Belsize Veg Bag scheme: The role of ICTs in enabling new voices and community alliances around local food production and consumption. *Futures*, 62(part B): 173–180. Available at: www.sciencedirect.com/science/article/pii/S0016328714000706 (accessed 20 February 2017).

Wilk, R. (2006) From wild weeds to artisanal cheese. In R. Wilk (ed.) *Fast Food/Slow Food: The Cultural Economy of the Global Food System*. Plymouth, UK: Altamira Press, pp. 13–27.

7 Displacement, 'failure' and friction

Tactical interventions in the communication ecologies of anti-capitalist food activism

Eva Giraud

This chapter conceptualises the dynamics of digital food activism in an anti-capitalist context, and explores the specific frictions surrounding this form of protest. Anti-capitalism is a notoriously slippery term (Chatterton, 2010: 1205–1206), which can encompass everything from socialist political parties to anarchist networks, but the focus here is on movements which have greater affinity with the latter. More specifically, this chapter's emphasis is on autonomous movements: non-hierarchical, leaderless, groups of activists who seek to craft grassroots spaces and lifestyle practices that are alternatives to capitalism (Pickerill and Chatterton, 2006).[1] These autonomous anti-capitalist alternatives can take a number of forms: from local social centres that are run without state support (Hodkinson and Chatterton, 2006), to large-scale protest camps that demand the construction of entirely different infrastructures to support the everyday lives of campers, including temporary food provision (Giraud, 2013), sewerage systems (Starhawk, 2005) and communications networks (Feigenbaum et al., 2013). The distinct difficulty of digital food activism, when it is bound up with the praxis of autonomous movements, is that both the promotion of particular consumption practices and the engagement with digital media have, historically, had a contentious (if integral) role within the work of these protest movements.

Frictions between activists' values and practices arise, in particular, when articulating arguments to broader publics; a number of commentators have foregrounded how both consumption narratives themselves (Guthman, 2008; DiVito-Wilson, 2013) and the media technologies used to construct these narratives (Skeggs and Yuill, 2016) can descend into elitist lifestyle narratives or commodify dissent in ways that undercut anti-capitalist values. There is a danger, therefore, that digital food activism can play an ambivalent role in anti-capitalist contexts (see Lekakis, 2013: 116–143). Frictions can also exist in the forms of internal communications that are integral to the identity-construction of activist groups; indeed both food and particular communications arrangements have been associated with the emergence of informal hierarchies, privileging those who adopt diets such as veganism (Giraud, 2013, 2015a) or have particular technological know-how (Nunes, 2005; Boler and Phillips, 2015), for instance.

One of the most prominent and long-standing instances of digital food activism, the McInformation Network, is drawn upon here to flesh out the relationships between these overlapping forms of friction, using a combination of documentary analysis and participatory action research. First engaging with digital media in 1996 with the launch of the influential McSpotlight website, McInformation has worked to construct a digital counter-narrative about globalisation and food production for over 20 years. Its early use of the Internet was seen as indicative of computer-mediated communication's potential to support international solidarity networks of resistance against capitalism (Atton, 2002, 2003; Pickerill, 2003). Despite early academic optimism about McSpotlight, the campaign has faced difficulties in the wake of broader political changes and the popularisation of social media and the corresponding decline in activist-produced alternative media (Lievrouw, 2011; Giraud, 2014). To conceptualise some of the frictions central to McInformation, the chapter engages with work that has called for a shift away from analysing specific media platforms, to instead analyse the communication ecologies – or evolving relationships between the range of communications media – that are associated with particular forms of activism or specific social movements (e.g. Treré, 2012; Feigenbaum *et al.*, 2013; Mercea *et al.*, 2016; Treré and Mattoni, 2015).

This focus on tensions surrounding the consumption practices and technologies involved in the communication ecologies of anti-capitalist food activism, builds on an emergent body of work focused on identifying digital frictions (Fuller and Goffey, 2012; Ruiz, 2014). In this body of work, friction refers to the difficulties that are routinely negotiated in everyday engagements with media technologies, which range from human confusion to technological glitches. Though in practice frictions are often just treated as a nuisance that needs to be 'worked around' in order for technologies to fulfil their desired role, recent conceptual work has foregrounded them in order to challenge assumptions that communication technologies can smoothly translate neoliberal (or indeed any) political ideals by shaping users' behaviour in predictable ways. In addition to its specific focus on food activism, this chapter thus also feeds into broader debates (e.g. Giraud, 2015b; Shea *et al.*, 2015) about whether frictions between technologies, activists and issues hold potential for complicating theoretical narratives about the commodification of dissent.

The core argument put forward here is that, through identifying some of these frictions and (building on Barassi, 2015) how they are navigated through activists' everyday practices, it becomes possible to see forms of political agency that are undervalued in narratives of commodification (e.g. Dean, 2009, 2010). These manifestations of agency include processes of tinkering with media arrangements in ways that subtly alter the affordances of particular media or juxtaposing different communication platforms in order to alter the resonance of specific acts of communication; here it is argued that these acts should not be seen as transitory and thus ineffectual but understood as 'tactical interventions': practices that derive from actors in positions of structural inequality but whose work nonetheless mediates the infrastructures they are embroiled within in enduring ways. Taking a careful look at activists' tactical interventions helps to elucidate how entanglements between alternative participatory media and proprietary social media are negotiated, in contexts where alternative media platforms have increasingly been displaced by commercial alternatives in order to reach beyond the immediate activist community.

Situating McInformation

Broader context: autonomy in the UK

As touched on above, both the provision of food (in contexts such as social centres and protest camps) and activist media initiatives have held an uneasy position within anti-capitalist praxis due to being sites where particular ideological values are reproduced or contested within the everyday work of activist groups. Before focusing on McInformation itself, therefore, it is important to provide some political context to understand the reasons behind these tensions.

Autonomous activism is prefigurative in its attempts to realise anti-capitalist futures in a capitalist present, as opposed to using more hierarchical organised forms of political organisation as a means to an end (Maeckelbergh, 2011). This form of politics is also, by extension, messy, as it is constantly having to negotiate tensions between, for instance, material constraints on time and resources and prefigurative ideals (Brown and Pickerill, 2009; Mason, 2014), or striving for non-hierarchical forms of political organisation while ensuring that structures are in place to guard against the emergence of 'informal hierarchies' (Nunes, 2005). The other key feature of autonomous activism is that – despite these tensions – it is nonetheless hopeful (Pickerill, 2008), and tries to negotiate these tensions in practice, even if it does not always succeed. The prefigurative dimension of this activism is especially evident in both the development of alternative media (Downing *et al.*, 2001; Atton, 2002; Kulick, 2014), and the promotion of alternative consumption practices, such as the consumption of gleaned food, localism or boycotting of brands seen as unethical (Harris, 2009; DiVito-Wilson, 2013; Giraud, 2013, 2015b), which have generated debate about how to develop infrastructures that can best prefigure activists' aims.

Autonomous news network Indymedia, for instance, epitomises attempts to prefigure autonomous ideals through engaging with particular media technologies. After experiments with using digital media to document protest events in London at 1999;s Carnival Against Capital, the first Indymedia centre was launched later that year in Seattle to provide a platform for anti-capitalist perspectives at the high profile series of protests against the World Trade Organisation (Kahn and Kellner, 2004). A broader media network grew out of this initiative, which was composed of local news sites that published entirely user-generated stories with this local content then feeding into regional and national hub sites. Local Indymedia centres were supported by editorial collectives, who were often associated with autonomous social movements, but meetings were open to anyone who wanted to be involved with concerted attempts to reach

beyond the immediate activist context to involve local communities in the production of alternative media (Pickerill, 2007). Unlike commercial social media,
the criterion for being listed as an Indymedia Centre was that collectives had to
maximise participation in all aspects of the news production process, a principle
that also underpinned the wider network (with network-wide decisions made
through consensus via global email lists) (Garcelon, 2006). These values also
extended to the technologies used, which were again developed and run by activists who used open source software and frequently ran sites from local servers
(Pickard 2006a, 2006b). Principles of openness were similarly integral to the
production of content, with minimal editorial control over user-produced content
and the use of anonymous publishing to avoid repercussions for individuals who
posted about politically sensitive issues (Stringer, 2013).

At its peak, the global network gathered over 175 regional Indymedia Centres
across nine continental regions (Giraud, 2014: 426), with each centre publishing
local news while being linked together (ideologically and in organisational and
infrastructural terms) by the overarching Indymedia network. At the same time,
however, the network has been the focus of heated debate, with a number of
well-documented problems, from spam and informal hierarchies (Uzelman,
2011) to tensions between centres in different global contexts. Frenzel *et al.*
(2010), for instance, foreground how Indymedia centres in Africa were treated
with a lack of sensitivity by other centres in the network who criticised them for
failing to uphold horizontal ideals, due to using resources and building space
provided by more hierarchically organised NGOs. Consensus decision-making
processes also led to urgently needed centres in Cairo and El Paso taking overly
long to approve decisions and resulted in frustration for activists in difficult
political situations (Giraud, 2014: 428).

The network was thus accused of ceasing to be prefigurative due to treating a
particular form of open, consensus-driven organisation as a model of how radical
media organisation should be realised, which did not respond to local needs or
changes in the broader political environment (Wolfson, 2012). Structures that
were designed to embody anti-capitalist, radically democratic values, in other
words, were accused of becoming hierarchical as key actors in the network
sought to make members adhere to a particular conception of what horizontality
should 'look like' that did not always reflect local conditions.[2] Indymedia is not
unique in facing these tensions, with similar allegations of informal hierarchies
levelled at more recent media initiatives within the Occupy Movement, this time
due to clashes between the desire to tightly control the public image of the
movement and the movement's participatory ideals (Boler and Phillips, 2015).

Food, similarly, has generated friction around its communicative role(s) in autonomous activism; like media, food provision in autonomous spaces is a key site in
which activist ideals are negotiated in practice. Decisions about which food to
consume, and how to gather, prepare and coordinate its provision, have been a contentious topic for anti-capitalist activism, and tensions often have to be negotiated
due to conflicting values in play surrounding food ethics. In the UK, for instance,
veganism has been promoted by activists who connect anti-capitalist politics with

environmentalism and animal rights praxis (for examples, see Anon, 2004; Morgenmuffel, 2005; Dominick, 2008). Certain collectives have also gained prominence in the UK activist community for providing food and catering infrastructure, such as Veggies Catering Campaign, Nottingham, and the Anarchist Teapot, in Brighton (Anarchist Teapot, 2005; Smith, 2009). The organisation of these groups – like that of the autonomous spaces they help to construct – attempts to prefigure autonomous political values and ideals of sustainability, in terms of their sourcing of local produce and cooperative organisational structures for instance (see Veggies, 2016a). The mediating role of food within activist infrastructures also segues into its more conventional symbolic function in activist contexts, wherein certain foods are used to signify anti-capitalist resistance (e.g. Clark, 2004). Practices such as creating meals from 'gleaned' food (that has been gathered from supermarket foods discarded in bins), for instance, mark symbolic opposition to mass consumption and waste (Mitchell and Heynen, 2009; Heynen, 2010; Sbicca, 2013), and vegan diets are often used to express a critique of the agricultural-industrial complex (Portwood-Stacer, 2012). Other strands of anarchism, however, perceive the symbolic use of vegetarian and vegan food in these contexts as being animal rights concerns that are unrelated to their own critique of capitalism, and hence the promotion of veganism not as prefigurative but as a hierarchical imposition (for an overview of these debates, see Giraud, 2013).

Individually, therefore, both food and media uses have the capacity to be prefigurative but have also been the site of tensions; the McInformation campaign helps to illustrate how these different sets of friction relate to one another in the context of digital food activism.

McInformation itself

As perhaps the archetypal target for anti-capitalist criticism, in the 1980s McDonald's became the focus of a small campaign by the London Greenpeace activist group who developed a five-page Fact Sheet that drew together the broad range of pre-existing criticisms that social justice groups had been levelling at the corporation (Vidal, 1997), and made allegations related to everything from aggressive advertising strategies and the promotion of unhealthy food, to the widespread de-skilling of kitchen workers and rainforest deforestation in the production of beef. In 1990, these protests gathered momentum after two activists involved in producing the pamphlet – Helen Steel and Dave Morris – were sued by McDonald's. The so-called 'McLibel' case went on to become the longest court case in British legal history (from 28 June 1994 to 19 June 1997), with its implications conceptualised across a number of fields: from critical legal studies (Wolfson, 1999) to environmental geography (Pickerill, 2003). After Steel and Morris were forced to represent themselves (due to a lack of legal aid for libel defendants in the UK), the McLibel Support Campaign and subsequent McInformation Network were launched in solidarity to raise funds and maintain the campaign in the face of what was perceived by activists as corporate bullying (for overviews, see Vidal, 1997; Armstrong, 2005).

This solidarity campaign went on to gain popular cultural attention as a forerunner of the global social justice movement (e.g. Klein, 2000: 365–396) and the launch of McSpotlight in 1996, which was frequently referred to within media studies as demonstrative of the Internet's capacity to support protest groups (Atton, 2002; Downey and Fenton, 2003). McSpotlight was initially launched by a group of UK activists who saw the potential for digital media to support the McLibel Support Campaign by providing an uncensored platform for criticisms of the corporation. The site was run along non-hierarchical principles with activists working to share technological skills in order to overcome divisions in levels of expertise. As a 'pre-Web 2.0' website, however, unlike the radical-participatory media initiatives (such as Indymedia) that proceeded it, McSpotlight was primarily an informational rather than an interactive site, with a core group of activists effectively acting as gatekeepers for information (though concerted attempts were made to broaden the membership of this core group and include content that linked to activist groups in other global contexts) (Pickerill, 2003: 55–63). Due to being so closely attached to the court case, academic and media attention towards the McInformation campaign has waned, yet its evolution – from the launch of McSpotlight itself to more contemporary anti-McDonald's Twitter campaigns – provides a striking instance of how the media used in a campaign can shift over time in ways that require adaptation in order to negotiate frictions between activists' aims and the tools they use.

Methodology

The contexts of anti-McDonald's protest and autonomous social movements form the backdrop to the specific instance of activism focused on in this chapter: performative food protests in the UK city of Nottingham, which were a local response to developments in the McInformation Network by some of the groups involved in it from its 1980s origins. The focus here is on one particular grassroots group aligned with the network: Veggies, the aforementioned 'campaign caterer' who have played an important role in supporting autonomous protest in the UK. Founded in 1984, the catering collective is currently based in an autonomous social centre in Nottingham, and provides food for events such as protest marches and community festivals (Veggies, 2016a). In addition, Veggies have a mobile kitchen that can support mass catering over longer periods of time in contexts such as protest camps; in the past, for instance, they have provided food at anti-nuclear protests at Faslane military base, UK Camps for Climate Action, and anti-G8 and G20 mobilisations, in addition to animal rights events (Smith, 2009). Veggies' role in the McInformation campaign is especially significant due to the group producing their own succinct 'What's Wrong with McDonald's' pamphlet, which preceded but went on to supersede the five-page fact sheet as the main leaflet distributed by McInformation.

Discussions of Veggies' work are derived from participatory action research (PAR) that I engaged in over a five-year period (2006–2010) as part of my doctoral project, which examined the work of a range of contemporary animal rights groups and the communicative tactics used by these groups in articulating animal rights issues. In this chapter I further contextualise my primary observations by referring to longer histories of digital media-use in the campaign over a 20-year period (1996–2016), drawing on secondary literature and primary sources that include a range of printed and electronic materials produced directly by the campaign, such as pamphlets, newsletters, protest reports, websites, and social media accounts.

PAR is distinct from ethnography in entailing not solely observation but more active forms of participation, which contribute to the practical work undertaken by the groups at stake. In ethical terms this form of research aims to develop knowledge through being 'empathic and interactive rather than extractive and objective' (Pickerill and Chatterton, 2006: 732). This involves adopting a position in which the lines between activism and research become blurred (Pain, 2003: 652–653), due to being both a 'commentator on' and 'embedded participant within' the groups whose work is engaged with in the research (Pickerill and Chatterton, 2006: 732). Key to the conception of participation that underpins this method is the sharing of skills (Pain, 2003: 253–254), the co-production of knowledge (Ostrom, 1995; Borda, 2001) or the collaborative production of resources (Chatterton, 2008: 424–426).

For instance, my research was able to inform the development of communicative resources, which included: collaborating on website and pamphlet text; setting up an email account and list; documenting protest events online; and printing a booklet of cheap and accessible recipes. In addition, I helped to coordinate protest events by preparing and distributing food; facilitating cookery skill-share workshops and coordinating protest materials. I also shared my technical skills with the group in a more direct way, drawing on my experiences in grant applications and report writing, in order to apply for (and then meet the obligations of) a small grant from the Vegan Society, which had changed its bursary procedure in a way that made group members unsure how to approach it.

The level of participation involved in PAR makes no pretence at neutral observation 'from a distance', but as Pickerill and Chatterton note, this does not equate to being unreflexive or uncritical and can support the on-going processes of self-reflection and internal critique that already occur within activist contexts (2006: 732). In summary, while this form of research generates a range of conceptual and ethical issues that have to be navigated (for further discussion, see Banks *et al.*, 2013, 2014), it nonetheless offers a level of understanding of the difficulties and tensions faced by grassroots activist groups that could not be acquired otherwise. In this instance, observations garnered from participating in the group are especially informative in understanding how shifts in the broader media and cultural environment can generate tactical adaptation on the part of the campaigners, tactics that are often unnoticed or undervalued in conceptual contexts.

Debates about digital activism: initial optimism in theory and practice

As Pollyanna Ruiz (2014) notes, the difficulty for contemporary protest movements is that they are often polyvocal and composed of coalitions between groups with slightly different motivations, even if they share similar aims. These groups thus 'frequently find their polyvocal position difficult to articulate in an arena accustomed to a single and unified narrative' (ibid.: 1). When the original anti-McDonald's Fact Sheet was published in the 1980s, McDonald's was so resonant as the focus of protest because it offered a thread capable of weaving a coherent polyvocal narrative together; as Helen Steel notes in Franny Armstrong's *McLibel* documentary (2005), McDonald's acted as a symbol for a range of concerns related to food justice to which activist groups working in different contexts had drawn attention.

The medium through which this narrative was articulated, however, generated frictions that had to be negotiated. Certain media – for instance, paper pamphlets – might derive from a social movement setting wherein their content is decided dialogically through 'collective processes of media production' (Barassi, 2013: 58), as with London Greenpeace's original pamphlet. When a pamphlet is printed and distributed, however, it congeals as a narrative that shifts from prefiguring directly democratic ideals (where articulations of issues are developed through consensus and with those who are implicated with these issues) to having a more representational relationship with those whose concerns it seeks to articulate.

From such a perspective, the way in which other activist groups' polyvocal concerns were drawn together through the original Fact Sheet could be seen as privileging certain activist groups as spokespeople for other movements. This form of communication also has potential to reinforce hierarchies between would-be advocates and those they are speaking for (in the case of the Fact Sheet, the 'spoken for' were workers, children, animals and a myriad other actors whose plight was being articulated by the pamphlet). A section on the original Fact Sheet entitled 'trained to sweat', for instance, stated that:

> It's obvious that all large chain-stores and junk-food giants depend for their fat profits on the labour of young people. McDonald's is no exception: three-quarters of its workers are under 21. The production-line system deskills the work itself: [a]nybody can grill a hamburger, and cleaning toilets or smiling at customers needs no training. So there is no need to employ chefs or qualified staff – just anybody prepared to work for low wages.

While this narrative offers a structural critique of de-skilling, it does so by positioning the workers in particular ways (as unskilled, doing jobs anyone can do, desperate for work and, crucially, as needing to be spoken for). During the McLibel trial a huge volume of evidence was amassed to support these claims, and 37 workers provided witness statements at the trial in support of the pamphlet; the trial, in other words, provided a platform for workers to speak for themselves.[3] Post-trial, however, some of the printed documents (particularly the Veggies' 'What's Wrong with McDonald's' pamphlet) were still regularly distributed, with their on-going distribution seen as an iconic symbol of resistance against McDonald's. In my own engagement with the campaign, the danger of particular documents establishing normative definitions of issues was highlighted at a local picket for the 'International Day of Action against McDonald's' (16 October 2007), where an ex-employee voiced her discomfort to us about the pamphlet's characterisation of workers as having no agency and being 'forced to smile'.

Digital media have frequently been framed as ameliorating some of these hierarchical issues. This deterministic line of argument was put forward, most famously, in Manuel Castells's work (e.g. 1997, 2012), but such claims have resonated in media studies texts that have praised the openness and visibility afforded by digital media (Allan, 2005; Gilmor, 2006). Even more nuanced analyses have drawn similar conclusions about certain technologies having attributes that intrinsically lend themselves to more democratic modes of organisation (e.g. Hands, 2010). As Rodrigo Nunes (2005) argues, assumptions about the capacity of particular media to guarantee 'openness' or flatten hierarchies have also persisted in activist contexts. Such narratives framed how McSpotlight was initially understood; when the website was launched in 1996, it was seen to exemplify digital media's potential to act as a tool for activists in 'David and Goliath' struggles between campaigners and multinational corporations (Downey and Fenton, 2003: 196). In early examinations of the Internet's capacity to support the work of protest movements, furthermore, McSpotlight was regularly cited to illustrate how digital media could be used to construct counter-narratives to gain visibility in mass media contexts and overcome corporate attempts to silence critics (Atton, 2002: 144–150).

Early analyses of the role of digital media for activists also stressed their 'deterritorialising' capacity in enabling activists to overcome local publishing restrictions (from legal constraints to economic overheads) and make material accessible to global audiences (e.g. Bach and Stark, 2004). Such arguments were again pivotal to narratives about McSpotlight which emphasised its value in overcoming the libel laws that had led the McLibel court case in the first place (Klein, 2000: 365; Pickerill, 2003: 11). Yet McSpotlight was not simply a means of overcoming the 'limitations' of paper pamphlets; equally important was the site's capacity to 'reterritorialise' the broad critique it made of McDonald's (and other multinational corporations), through supporting grassroots campaigns in which local groups drew attention to the specific ways in which the broad set of concerns articulated by McSpotlight impacted upon their local area.

An integral part of McSpotlight was not just acting as an online archive (with activists aware – even in the relatively early stages of digital media use – that web-based resources were likely to preach to the converted) (Pickerill, 2003: 48–49), but providing resources for activists to conduct their own grassroots campaigns. To facilitate this, the site was home to a campaign called 'Adopt-a-Store', which encouraged local groups to print off and distribute their own pamphlets for local pickets. The campaign thus combined online and offline media, in order to appeal to different audiences, with McSpotlight acting as a hub to support autonomous praxis without acting in a hierarchical manner and dictating how local groups should realise protest actions (Atton, 2002).

Shifting uses of media; narratives of failure and displacement

The trajectory of McInformation, however, parallels shifts in both the use and the conceptualisation of activist engagements with digital media. Just as high-profile online alternative media initiatives have struggled and been displaced by social media (Lievrouw, 2011), initial optimism about the capacity of digital media to support the work of activists has been displaced by perspectives that see these media as commodifying communication (Dean, 2010; Juris, 2012). As explored in further depth below, these narratives are overly neat, but they do point to some dominant trends, as reflected by shifts in anti-McDonald's campaigning.

As Jenny Pickerill noted, even at the peak of McSpotlight it was clear that the site would be transitory due to its attachment to a particular event (2003: 8); indeed, though the site remains as an archive of the trial, the last time it had a substantive update was in 2003 and the forums were waning in popularity by 2002. These shifts in the media environment were visible locally in my own research, in relation to the role of different media used in the practical work of activists associated with Veggies, whose campaigning work is informative for illustrating the forms of friction and negotiation that occur on a grassroots level.

My first experience with the campaign (15 October 2006) involved handing out leaflets to mark the World Day of Action against McDonald's, at a small local picket outside a McDonald's in Nottingham city centre. While there was little engagement with McSpotlight, other than mentioning it to people as we were handing out leaflets, Indymedia had a visible role in promoting the event in advance and subsequently documenting it. A local activist arrived and took photographs of the protest, for instance, which were subsequently written into a report that included further links to McSpotlight (Lodge, 2006). Displaced by the more participatory Indymedia, McSpotlight's role shifted into having an archival function (in addition to being referred to by activists at the picket as a source of evidence, it was also linked to on the Indymedia report for this purpose), while Indymedia itself offered a more responsive publishing platform in enabling activists to rapidly self-publish information rather than going through any form of gatekeeper.

Just as McSpotlight was displaced by Indymedia, however, Indymedia itself was eventually displaced by social media. Yet it is important to note that engagements with social media were not uncritical; even in 2010 when social media was being discussed positively in conceptual contexts (Gilmor, 2006; Castells, 2012), there was wariness on a grassroots level. The presence of a police car at a local autonomous social centre, which arrived prior to a the UK Spring Animal Rights gathering (13 March 2010), for instance, led to debate in the Nottingham activist community about whether caution should be taken about promoting events on social media; the unusually young demographic of the activists who attended, moreover, was also reflected on amidst concerns that a reliance on social media could be exclusionary of older demographics of activists. In present-day campaigning work, however, social media is an everyday part of activists' communication repertoires: Veggies's website, for instance, now embeds their Twitter feed, and recent anti-McDonald's protests have been promoted with hashtags that articulate these actions in relation to other campaigns to create affinities (a trend discussed in more depth below).

These local developments resonate with theoretical narratives, which have pointed to the 'failures' of key radical-participatory online media initiatives (Lovink and Rossiter, 2009), in the face of social media. In Jodi Dean's terms, these trends reflect a series of 'displacements' wherein a particular communications technology 'triggers a process of change even as change quickly overtakes it' (2010: 26) and 'media themselves may be displaced by the events and developments they enable' (ibid.: 28). In this instance, it seems as though radical-participatory platforms that have mediated protest in distinct ways have been successively displaced by other – commercial – media that commodify participation. Whereas Indymedia and McSpotlight were produced by and prefigurative of the communities they were representing, the settings of commercial social media (particularly Facebook and Twitter) have been criticised for their implication in processes of data-mining and self-commodification (Skeggs and Yuill, 2016). Social media have also been seen to promote a politics based not on sustained processes of dialogue and consensus – which are necessary to support processes of network-building and coordinated action – but instead offer more fragmentary processes of 'aggregation' (Juris, 2012), where individuals are brought together in much looser arrangements that often lack the capacity to cohere as concrete forms of praxis (Dean, 2012).

On one level, therefore, what emerges when examining the changing uses of media in McInformation is a story of failure and displacement. The successive 'failure' of activist media initiatives, in the face of Web 2.0 media, has effectively led to the commodification of dissent, with social media use 'perpetually expanding the topography of struggle even as it constantly signals the locations, intentions and networks of those who are fighting' (Dean, 2010: 125). If these technologies are treated synonymously within activist praxis, moreover, Dean argues that there is the danger that the setting of technologies (and the commercial logics that structure them) will become ignored, as will their capacity to undermine dissent (ibid.: 28).

Yet, both 'pessimistic' and 'optimistic' narratives adopt a somewhat deterministic position, which neglects the messiness of activists' material engagements with digital media, and all the frictions and unexpected manifestations of agency that go with it. It is these frictions that are revealed when focusing on more recent developments in the McInformation campaign, and when the focus is shifted away from specific media platforms and towards a more ecological perspective of their campaign.

A move towards ecologies

Though much early media research stressed the value of McSpotlight in gaining visibility, acting as a radical resource and supporting campaigns, a small number of analyses hinted at the direction that the study of social movement media was to take in the late 2010s. Analyses that derived from a social movement studies perspective, for instance, emphasised how McSpotlight was 'just one aspect of a varied strategy co-ordinated by the McLibel Support Campaign', among other media and political tactics including 'mass leafleting, media focus, pickets outside McDonald's stores and international days of action, with links to residents' opposition groups and disgruntled McDonald's workers' (Pickerill, 2003: 10). This account is resonant of more contemporary research, which has taken an ecological focus wherein specific technological platforms have not been focused on in isolation but situated within wider (and constantly evolving) media arrangements and repertoires of protest tactics (e.g. Treré, 2012).

Recent developments in ecological approaches have departed from deterministic roots in medium theory (Mercea *et al.*, 2016: 280) by offering a more sustained focus on the entanglements between, for instance, activist cultures, media technologies and protest actors. Although the affordances of different media are emphasised, these affordances are not understood as static technical attributes; instead, within contemporary communications research the affordances of 'each individual medium [are] defined in relational terms in the context of all other media' (Madianou and Miller, 2013: 170). In addition to particular platforms being defined through their relationships with other media technologies, 'the cultural, somatic, technical and historical cannot be separated into distinct and discreet categories because the material components and thresholds of an object are intimately related to their design, manufacture and use' (Ash, 2014: 85). When media assemblages are described as being composed of entangled relations (e.g. Feigenbaum *et al.*, 2013; Feigenbaum, 2014; Boler and Phillips, 2015; Shea *et al.*, 2015), therefore, this does not just infer that disparate agencies become tangled with one another in protest contexts, but is also more akin to the new materialist assertion that entities gain a discreet form by and through their relationships with one another (in line with Barad, 2007). Entanglement thus offers a messier account of the multitude of agencies in play within particular media arrangements, as opposed to notions of hybridity or co-constitution that – as Jamie Lorimer notes (2015: 24) – can infer only two co-shaping partners.

What is valuable about understanding communication in ecological terms, for the study of anti-capitalist activism in particular, is that such approaches lend themselves to an exploration of how the affordances of particular media technologies can prefigure non-hierarchical modes of politics. Unlike the celebratory and critical narratives discussed above, however, ecological frameworks resist technologically or culturally deterministic readings. Ecologies offer, in other words, 'a means to ground the empirical verification of the degree and measure the chances (socio-economic, cultural and political) to which the participatory ideal may be attained in the messy interconnections among media, users and their manifold local settings' (Mercea *et al.*, 2016: 281). The participatory affordances of social media, for example, cannot be assessed through focusing on apparently rigid technical features, or by examining its setting in communicative capitalism. Instead, attention must be paid to how technologies are used and engaged with in practice, as opposed to in idealised ways (Couldry, 2012). It is important, in other words, to examine how tensions between technologies and political aims are negotiated in practice by specific groups and individuals (Barassi, 2015).

What this means in a more concrete sense is illustrated through focusing on the Veggies' website, which complicates the narratives of failure and displacement that have characterised critiques of activist uses of social media. The site has a specific page dedicated to food activism (Veggies, 2016b; Figure 7.1), which contains images of past food give-aways; a feed featuring listings from Veggies' own diary of food activism events in the UK; leaflets about food give-aways that can be printed for distribution; downloadable legal briefing documents; links to past give-aways on Indymedia; and tweets embedded from their live Twitter feed. Juxtapositions of 'old' and 'new' platforms are often characterised as part of the mundane aesthetics of digital media (Fish, 2015: 90), but what should be emphasised in this instance is the *political* dimension of this process as juxtaposition becomes part of the activists' memory-work and legacies, with newer initiatives articulated as part of a longer history of campaigning.

Figure 7.1 Screenshot of food giveaway advice and reports from Veggies' website: www.veggies.org.uk/campaigns/vegan/free-food-givaways.

Indymedia, for instance, has not been displaced wholesale by social media, but is juxtaposed with it, with links to old Indymedia reports about Nottingham food activism listed next to the live Twitter feed that relays recent campaigning work. Indymedia still exists, therefore, but its affordances have changed due to shifts in activist media practices and the broader media ecological context, and the network now fulfils an archival role rather than acting as a participatory platform that is used for mobilisation or to document unfolding events.

Twitter, conversely, has its affordances subtly mediated through its own juxtaposition with activist-produced alternative media. The network has been characterised as the archetypal tool for communicative capitalism in its aggregative logic, which encourages individualistic forms of self-promotion and can only foster weak social ties (Dean, 2010; Juris, 2012). This perceived logic of Twitter, in turn, means the platform lends itself to ephemerality (Murthy, 2011), and the capacity to speak and rapidly disseminate messages: 'fragments thought into ever smaller bits, bits that can be distributed and sampled, even ingested and enjoyed, but that in the glut of multiple, circulating contributions tend to resist recombination into longer, more demanding theories' (Dean, 2010: 2). These affordances, however, are given new resonance in the context of Veggies' website; micro-blogs that are seemingly self-promotional are articulated in relation to the group's sustained work with radical social movements in the UK. Tweets such as #VeganOutreach, for instance, could be seen as the promotion of ethical lifestyle politics, but this connotation is ameliorated by its contextualisation on Veggies' website where it appears below the feed from their self-produced Campaign Diary, which features more detailed information about the anti-capitalist settings in which this outreach takes place and includes events such as climate marches and a local critical mass bike ride. The aforementioned Indymedia reports, similarly, offer a means of situating these tweets in relation to a longer history of campaigning work. The affordances of the 'hashtag politics' used by the group – a tactic that has been criticised for crystallising the forms of self-promotion and liberal individualism characteristic of communicative capitalism – are thus given a different resonance if understood as part of a more complex communication ecology.

From this lens what emerges is not a narrative of failure and displacement but of an evolving communication ecology where the affordances of media are entangled with their broader socio-technical context. What is also revealed are the processes of negotiation that occur when activists engage with commercial social media (as with Barassi, 2015), in this instance, these negotiations involve a re-contextualisation of social media through its juxtaposition with more radical media; ephemerality and aggregative logics, in other words, are translated into a more sustained political identity. This focus also reveals scope for activists to 'tinker' with the affordances of particular media, by intervening in these ecologies in tactical ways. This form of tinkering was evidenced, in particular, through the increased use of food in local protests from 2006, to create more inclusive dialogical spaces within activists' communication ecologies.

The 'other media' of (digital) food activism: food itself

Although digital media serve an important role in movement-building, so too do the more mundane actors, which constitute the lived environment of protest camps, social centres and other spaces engaged with by autonomous activism in the UK. In the context of the protests at stake here, if – as Anna Feigenbaum (2014) suggests – further attention is paid to the 'other media' of activism, which go beyond those conventionally understood as media technologies, then the important mediating role of food itself is made explicit.

'Other media' are actors that adopt communicative affordances when they are part of specific socio-technical assemblages. The mediating capacity of objects is not restricted to their symbolic use, moreover; as Feigenbaum argues, while actors such as tents can – in the context of protest camps – articulate as well as enact occupation, the difficulty, complexity and material work involved in constructing alternative infrastructures also generate dialogue that 'can help clarify what is at stake to whom in these claims' of occupation (2014: 19). In addition, 'affect and emotion are bound up in object encounters, so too are tactical knowledges and embodied experience' (ibid.: 17); like more conventional media technologies, therefore, the objects used to construct autonomous infrastructures mediate not only action but also communication in specific ways, and are hence constitutive of how prefigurative politics is realised and expressed.

Including 'other media' in discussions of activist media ecologies is especially valuable for food activism, due to the significance of these media for broader activist infrastructures; simply put, in order to use food in protest, the necessary infrastructures have to be in place to produce and distribute it. The frictions generated when activists discuss and reach consensus over how these infrastructures are developed – as with Feigenbaum's tent example – offer a productive space for sharing expertise and reaching mutual understanding. The digital media used by Veggies in promoting and documenting their food giveaways, for instance, go beyond the forms of 're-territorialisation' offered by McSpotlight. These media do not just act as a hub for a pamphleteering network, but offer advice on crafting the infrastructures necessary for local groups to develop their own grassroots protests (offering links to resources and their own advice sheets on the legalities of this sort of protest).

These media and food-provision infrastructures are further entangled, if understood in line with Veronica Barassi's (2013) argument that media arrangements are intimately bound up with the production and organisation of activist spaces. As with other recent work in media theory, which has drawn inspiration from science studies or new materialism (such as the aforementioned accounts of entanglement), Barassi emphasises that activist networks are not rigid structures that mirror the fixed architectures of media technologies, as with work that has argued network technologies facilitate networked social movements (Castells, 1997; Hands, 2010). Engaging with Latour, Barassi instead understands networks as emerging *through* particular sets of associations (2013: 51); this perspective thus moves beyond determinism while still foregrounding the role of particular media arrangements in the production of activist spaces and social organisation. This attention to the spaces that are produced is especially important for activism that is bound up with food provision, because particular communication ecologies enact a mode of food production and consumption that constructs and privileges certain spaces as well as organising the relations in these spaces.

As Anastasia Kavada (2015) points out, different media platforms work together to engage with different audiences and hence overcome one another's limitations. Veggies' digital media ecologies, for instance, involved email lists (which reinforced group identity); websites and online alternative media (which engaged with the wider activist community to publicise events); and social media (to engage with those with 'weaker ties' to the activist movement, or foster more such ties). At the same time as working collectively to increase accessibility, however, these media arrangements simultaneously produced and privileged certain groups and spaces; the local animal rights group and Veggies catering campaign, for instance, had pre-existing social media and their own website, respectively, which could be used to publicise events effectively, but this had the side effect of establishing them as privileged nodes.

The infrastructures of food production itself were also bound up with these arrangements, as the privileging of certain media also affected who was included and involved in the spaces of activism. The need to produce and distribute food meant that it seemed 'common sense' to use existing infrastructures – such as Veggies' kitchen – as the space where this occurred. Skill-share workshops designed to involve members of the public, similarly, took place in spaces available to the autonomous groups initiating the protest. Although a range of people attended these workshops who were not directly related to the animal rights group or Veggies catering campaign, they did tend to have ties with the broader activist community, which were reinforced by a reliance on activists' pre-existing media arrangements.

It is important, however, to recognise that this privileging of particular spaces and groups through the dynamics of activist communication ecologies was not naïve or uncontested. Indeed, the use of food in the first place was – in part – an attempt to move beyond the more didactic approach of pamphleteering, or media use whose audience had pre-existing ties to activist groups, in order to open space for further dialogue with would-be McDonald's consumers (see Giraud, 2015a, for further reflection). In line with work adopting a relational understanding of the affordances of different technologies, the affordances of food can be seen as mediating protest in distinct ways when used as a communicative tool in this context. Unlike the Big Macs consumed inside McDonald's, food such as veggie burgers and soya milkshakes were given certain communicative affordances through being distributed in public space, for free, and being accompanied by activist literature and activists themselves.

Like protest tents, the performative act of distributing free food in public space articulates a critical message about consumption practice. This message can vary depending on its specific protest setting; as Nik Heynen (2010) notes, for instance, groups such as Food Not Bombs have long used food distribution to highlight food poverty and contest regulations that exclude homeless populations from public space. In the context of the McInformation campaign, the distribution of food offered a more specific critique of McDonald's but one that actively involved the public in this articulation through disrupting existing consumption patterns and by creating space for dialogue. Understanding food in this way is important, because it offers space to understand how 'other media' can be tactically used to intervene in existing communication ecologies, in ways that affect the 'environment of affordances' that characterise these ecologies (Madianou and Miller, 2013: 170). In the protests I was involved with, for instance, while this dialogical space fostered dialogue with a range of different publics – from suspicious teenagers to those with a pre-existing interest in food politics – it also created space for these publics to discuss the campaign with activists, in ways that foregrounded informal hierarchies and exclusions that may not have been visible otherwise.

Conclusion

What has emerged through focusing on different iterations of McInformation – looking at the campaign's development over time, examining a range of online and offline documents and foregrounding tactics that arose at a grassroots level – are the specific frictions associated with digital food activism in anti-capitalist contexts. Tensions that have long existed surrounding both the role of food and the role of digital media overlap in informative ways within the campaign, and highlight the barriers to realising prefigurative ideals through both consumption practices and media use. What is also foregrounded, however, especially through a focus on the grassroots level, is the need to avoid deterministic assessments about the role of different media in these campaigns (whether these assessments are celebratory or grounded in critique). Instead, through taking an ecological approach and developing a more relational understanding of the affordances of media technologies (an approach that allows food itself to be understood as a communications platform), it becomes possible to see how particular technological arrangements, practices and tactics can subtly re-shape the affordances of media to support activists' prefigurative work and navigate some of these frictions.

The story of anti-McDonald's campaigning thus holds particular value for understanding the dynamics of digital food activism. Grassroots actions such as the food give-aways in which Veggies engaged, for instance, have long been used by anti-capitalist groups and are useful in directly intervening in everyday contexts of consumption. Yet, although these interventional tactics have an immediate impact, platforms such as McSpotlight and Indymedia have historically proven important in transforming local interventions into more visible, scale-able and durable anti-capitalist counter-narratives, in ways that can support future protest. By turning attention to more recent work within anti-McDonald's campaigns, however, it is possible to see how mutually supportive relationships between interventional protest and digital food activism can become complicated by a shifting media environment. In a context where it is often necessary to engage with proprietary media platforms in order to sustain engagement with would-be consumers, grassroots actions have fed back into the digital media environment in order to maintain an oppositional message. The campaign foregrounds, therefore, the value of developing situated and historically-informed understandings of the entanglement of 'offline' and digital food activism, rather than focusing on specific media platforms at particular moments in time.

Activist tactics – both online and offline – are commonly understood as tactics in Michel de Certeau's sense (1984), in that it is not always possible for grassroots anti-capitalist activism to construct entirely autonomous spaces. Instead, activism often involves working within existing systems and structures rather than transforming them. This analysis has been especially important for work in media studies, with recent research into anti-capitalist protest foregrounding how activists are constantly negotiating tensions between their political aims and the commercial dynamics of digital media (Barassi, 2015; Shea *et al.*, 2015). Focusing on the digital food activism central to McInformation highlights a more sustained dimension to activist tactics, as processes of tinkering and tactical engagements with 'other media' can mediate broader communication ecologies in ways that have enduring implications for how activism is articulated and enacted. It is suggested, therefore, that instead of emphasising processes of failure and displacement, which focus on the transience of digital media, the process of navigating some of these frictions could be framed in terms of 'tactical interventions'. This concept acknowledges the constraints of working within existing systems but also foregrounds the endurance of activists' mediated agency.

Notes

1 The label 'autonomous anti-capitalist activism' is derived from how particular activists have characterised their own work, and marks purposive attempts by social movements to craft alternative infrastructures and ways of organising that are underpinned by values of social justice. The social movement origins of this concept and its normative political commitments to equality are thus distinct from more recent theoretical characterisations of processes such as digital sharing as marking 'post-capitalist' economies (e.g. Mason, 2015).

2 This idea of particular conceptions of non-hierarchical political organisation becoming normative models for what radical democracy should 'look like' is taken from Rodrigo Nunes's (2005) informative reflections on the work of the anti-G8 Dissent Network, prior to the 2005 G8 Summit at Gleneagles.

3 A practice continued on the (now defunct) McSpotlight forums, which had a section dedicated to workers.

References

Allan, S. (2006) *Online News*. Maidenhead: Open University Press.

Anarchist Teapot (2005) *Feeding the Masses*. Brighton: Anarchist Teapot.

Anon (2004) *Beasts of Burden*. London: In the Spirit of Emma c/o Active Distribution.

Armstrong, F. (dir.) (2005) *McLibel*. London: Spanner Films.

Ash, J. (2014) Technology and affect: Towards a theory of inorganically organised objects. *Emotion, Space and Society*, 14(1): 84–90.

Atton, C. (2002) *Alternative Media*. London: Sage.

Atton, C. (2003) Reshaping social movement media for a new millennium. *Social Movement Studies*, 2(1): 3–15.

Barad, K. (2007) *Meeting the Universe Halfway*. Durham, NC: Duke University Press.

Barassi, V. (2013) Ethnographic cartographies: Social movements, alternative media and the space of networks. *Social Movement Studies*, 12(1): 48–62.

Barassi, V. (2015) *Activism on the Web*. London: Routledge.

Boler, M. and Phillips, J. (2015) Entanglements with media and technologies in the Occupy Movement. *Fibreculture*, 26: 236–267.

Brown, G. and Pickerill, J. (2009) Space for emotion in the spaces of activism. *Emotion, Space and Society*, 2(1): 24–35.

Castells, M. (1997) *The Power of Identity*. Bodmin: Blackwell.

Castells, M. (2012) *Networks of Outrage and Hope*. Cambridge: Polity.

Chatterton, P. and Pickerill, J. (2010) Everyday activism and the transitions towards post-capitalist worlds. *Transactions of the Institute of British Geographers*, 35(4): 475–490.

Clark, D. (20040 The raw and the rotten: punk cuisine. *Ethnology*, 43(1): 19–31.

Couldry, N. (2012) *Media, Society, World*. London: Polity.

Dean, J. (2010) *Blog Theory*. Bodmin: Polity.

Dean, J. (2012) *The Communist Horizon*. London: Verso.

De Certeau, M. (1984) *The Practice of Everyday Life*. Berkeley, CA: University of California Press.

DiVito-Wilson, A. (2013) Beyond alternative: Exploring the potential for autonomous food spaces. *Antipode*, 45(3): 719–737.

Downey, J. and Fenton, N. (2003) New media, counter-publicity and the public sphere. *New Media & Society*, 5(2): 185–202.

Downing, J., Villarreal Ford, T., Gil, G. and Laura, S. (2001) *Radical Media*. Thousand Oaks, CA: Sage.

Feigenbaum, A. (2014) Resistant matters: Tear gas, tents and the 'other media' of Occupy. *Communication and Critical/Cultural Studies*, 11(1): 15–24.

Feigenbaum, A., Frenzel, F. and McCurdy, P. (2013) *Protest Camps*. London: Zed Books.

Fish, A. (2015) Mirroring the videos of anonymous: Cloud activism, living networks and political mimesis. *Fibreculture*, 26: 188–207.

Frenzel, F., Böhm, S. and Quinton, P. (2010) Comparing alternative media in north and south. *Environment and Planning A*, 43(5): 1173–1189.

Fuller, M. and Goffey, A. (2012) *Evil Media*. Cambridge, MA: MIT Press.

Garcelon, M. (2006) The Indymedia experiment. *Convergence*, 12(1): 55–82.

Gilmor, D. (2006) *We Are Media*. Sebastopol: O'Reilly.

Giraud, E. (2013) Beasts of burden: Productive tensions between Haraway and radical animal rights. *Theory, Culture and Critique*, 54(1): 102–120.

Giraud, E. (2014) Has radical participatory media really 'failed'? Indymedia and its legacies. *Convergence*, 20(4): 419–437.

Giraud, E. (2015a) Practice as theory: Learning from food activism and performative protest. In R. Collard and K. Gillespie (eds) *Critical Animal Geographies*. New York: Routledge.

Giraud, E. (2015b) Subjectivity 2.0: Digital technologies, pervasive drama and communicative capitalism. *Subjectivity*, 8(2): 93–179.

Guthman, J. (2008) Bringing good food to others: Investigating the subjects of alternative food practice. *Cultural Geographies*, 15(4): 431–447.

Hands, J. (2010) *@ Is For Activism*. London: Pluto Press.

Harris, E. M. (2009) Neoliberal subjectivities or a politics of the possible? Reading for difference in alternative food networks. *Area*, 41(1): 55–63.

Heynen, N. (2010) Cooking up non-violent civil-disobedient direct action for the hungry. *Urban Studies*, 47(6): 1225–1240.

Hodkinson, S. and Chatterton, P. (2006) Autonomy in the city: Reflections on the social centres movement in the UK. *City*, 10(3): 305–315.

Juris, J. (2012) Reflections on #occupy everywhere: Social media, public space and emerging logics of aggregation. *American Ethnologist*, 39(2): 259–279.

Kahn, R. and Kellner, D. (2004) New media and internet activism: From the 'Battle of Seattle' to blogging. *New Media & Society*, 6(1): 87–95.

Kavada, A. (2015) Creating the collective: Social media, the Occupy Movement and its constitution as a collective actor. *Information, Communication & Society*, 18(8): 872–886.

Klein, N. (2000) *No Logo*. London: Flamingo.

Kulick, R. (2014) Making media work for themselves: Strategic dilemmas of prefigurative work in independent media outlets. *Social Movement Studies*, 13(3): 365–380.

Lekakis, E. (2013) *Coffee Activism and the Politics of Fair Trade and Ethical Consumption in the Global North*. Basingstoke: Palgrave Macmillan.

Lievrouw, L. (2011) *Alternative and Activist New Media*. London: Polity.

Lodge, A. (2006) Nottingham's part of the Worldwide Day of Action against McDonald's. *Indymedia UK*. Available at: https://indymedia.org.uk/en/2006/10/353559.html (accessed 15 October).

Lorimer, J. (2015) *Wildlife in the Anthropocene*. Minneapolis, MN: University of Minnesota Press.

Lovink, G. and Rossiter, N. (2009) The digital given: 10 Web 2.0 theses. *Fibreculture*, 4. Available at: http://fourteen.fibreculturejournal.org/fcj-096-the-digital-given-10-web-2-0-theses/ (accessed 8 July 2013).

Madianou, M. and Miller, D. (2013) Polymedia: Towards a new theory of digital media in interpersonal communication. *International Journal of Cultural Studies*, 16(2): 169–187.

Maeckelbergh, M. (2011) Doing is believing: Prefiguration as strategic practice in the alterglobalization movement. *Social Movement Studies*, 10(1): 1–20.

Mason, K. (2014) Becoming citizen green: Prefigurative politics, autonomous geographies, and hoping against hope. *Environmental Politics*, 23(1): 140–158.

Mason, P. (2015) *Postcapitalism*. London: Penguin.

Mercea, D, Ianelli, L and Loader, B. D. (2016) Protest communication ecologies. *Information, Communication & Society*, 19(3): 279–289.

Mitchell, D. and Heynen, N. (2009) The geography of survival and the right to the city. *Urban Geography*, 30(6): 611–632.

Murthy, D. (2011) Twitter: Microphone for the masses? *New Media & Society*, 33(5): 779–789.

Nunes, R. (2005) Nothing is what democracy looks like: Openness, horizontality and the movement of movements. In D. Harvie, K. Milburn, B. Trott and D. Watts (eds) *Shut Them Down!* Leeds and Brooklyn: Dissent! and Autonomedia, pp. 299–319.

Pickard, V. (2006a) Assessing the radical democracy of Indymedia. *Critical Studies in Media Communication*, 23(1): 19–38.

Pickard, V. (2006b) United yet autonomous: Indymedia and the struggle to maintain a radical democratic network. *Media, Culture & Society*, 28(3): 315–336.

Pickerill, J. (2003) *Cyberprotest: Environmental Activism On-line*. Manchester: Manchester University Press.

Pickerill, J. (2007) Autonomy on-line: Indymedia and practices of alter-globalisation. *Environment and Planning A*, 39(11): 2668–2684.

Pickerill, J. (2008) The surprising sense of hope. *Antipode*, 40(3): 482–487.

Portwood-Stacer, L. (2012) Anti-consumption as tactical resistance: Anarchists, subculture and activist strategy. *Journal of Consumer Culture*, 12(1): 87–105.

Ruiz, P. (2014) *Articulating Dissent*. London: Pluto.

Sbicca, J. (2013) The need to feed: Urban metabolic struggles of actually existing radical projects. *Critical Sociology*, 40(6): 817–834.

Shea, P., Notley, T. and Burgess, J. (2015) Editorial: Entanglements – activism and technology. *Fibreculture*, 26: 1–6.

Skeggs, B. and Yuill, S. (2016) Capital experimentation with person/a formation: How Facebook's monetization refigures the relationship between property, personhood and protest. *Information, Communication & Society*, 19(3): 380–396.

Smith, P. (2009) Burgers not bombs. *Peace News*, 2514.

Starhawk (2005) Diary of a compost toilet queen. In D. Harvie, K. Milburn, B. Trott and D. Watts (eds) *Shut Them Down!* Leeds and Brooklyn: Dissent! and Autonomedia, pp. 185–201.

Stringer, T. (2013) This is what democracy looked like. In J. Juris and A. Khasnabish (eds) *Insurgent Encounters: Transnational Activism, Ethnography and the Political*. Durham, NC: Duke University Press, pp. 318–341.

Treré E. (2012) Social movements as information ecologies. *International Journal of Communication*, 6: 2359–2377. Available at: http://ijoc.org/index.php/ijoc/article/view/1681 (accessed 6 April 2014).

Treré, E. and Mattoni, A. (2016) Media ecologies and protest movements: Main perspectives and key lessons. *Information, Communication & Society*, 19(3): 290–306.

Uzelman, S. (2011) Media commons and the sad decline of Indymedia. *The Communication Review*, 14(4): 279–299.

Veggies (2016a) Veggies catering campaign. Available from: www.veggies.org.uk/about/ (accessed 16 May 2016).

Veggies (2016b) Free vegan food give-aways. Available from: www.veggies.org.uk/campaigns/vegan/free-food-givaways/ (accessed 16 May 2016).

Vidal, J. (1997) *McLibel: Burger Culture on Trial*. Chatham: Pan Books.

Wolfson, D. (1999) *The McLibel Case and Animal Rights*. London: Active Distribution.

Wolfson, T. (2012) From the Zapatistas to Indymedia: Dialectics and orthodoxy in contemporary social movements. *Communication, Culture & Critique*, 5: 149–170.

8 'Both fascinating and disturbing'

Consumer responses to 3D food printing and implications for food activism

Deborah Lupton and Bethaney Turner

Introduction

Food activism often involves challenging established modes of food production and consumption, which may mean looking for unusual alternatives to food production and promoting unfamiliar ingredients or modes of preparing and processing materials for human consumption as acceptable and valuable. In this chapter, we address the topic of 3D printed food – edible products fabricated using digital devices and software – and its implications for food activism. The production of 3D printed food has the potential to address some of the challenges targeted by food activists, including alleviating world hunger, improving food sustainability and reducing food waste. These possible uses of 3D printing technologies are currently in the development, experimental or speculative stage. These technologies offer the opportunity to examine how innovative modes of processing and preparing foods are greeted by members of the public. Discerning consumers' responses to emerging technologies when they are still largely little known and in the innovation phase of development can serve to explore how such responses are formed before familiarity develops, as well as helping to anticipate emerging social or political issues (Marcu *et al.*, 2015).

Very little research has been conducted thus far into what members of the public know about 3D food printing and their attitudes about consuming printed food products. In this chapter, we report some finding from our qualitative study into Australian consumers' attitudes and responses to the idea of 3D food printing. We begin with providing an overview of 3D food printing and how it has been discussed in relation to food activism causes. This is followed by consideration of some theoretical perspectives that provide insights into how consumers deal with novel food technologies and the new foodstuffs they produce. We then introduce our study and discuss findings that are pertinent to the possible uses of this digital technology in food activist causes.

3D food printing: an overview

Three-dimensional (3D) printing (more technically known as additive manufacturing) involves the use of computer-aided design (CAD) software instructing a digital fabricating machine to extrude materials in a layering pattern to form objects. 3D printing has been used for more than two decades in manufacturing (Petrick and Simpson, 2013) and is hailed in some quarters as a disruptive and revolutionary technology contributing to a third industrial revolution regarding the changes it may bring to manufacturing and the global supply chain and economy (Mohr and Khan, 2015). Companies and research labs began developing prototypes of 3D printing technology to create food products over a decade ago (Lipton *et al.*, 2015; Sun *et al.*, 2015). Cartridges filled with edible matter – pastes, purees, powders, doughs, liquids and gels made from substances such as sugar, chocolate, cheese, flour, fruit, vegetables and animal proteins – are used in this process. The pastes are extruded through nozzles to generate products, and can achieve intricate designs. Thus far, such objects as chocolate or sugar decorations, sweets and desserts, pancakes, pizza bases, biscuits, bread, pasta and shapes made from purees have been made using 3D printing technologies.

The fabrication of food using 3D printing technology has received some attention in the past few years; not only in the academic food industry and food science journals but also in news reports and blogs, particularly those focused on 3D printing in general (Lupton, in press). Since 2015, several 3D food printing conferences have been held in the Netherlands and Australia, organised by Jakajima, a company that focuses on fostering innovation. The conferences brought together entrepreneurs, developers, healthcare professionals, food industry representatives and university researchers working on the technology.

In these forums, it has been suggested that this technology has many uses. These may be grouped into the following categories:

- *Novelty/fun/creativity*: fine-dining and creative gastronomy; creative domestic cooking; decorative items (e.g. for cakes); food for special occasions; food for children.
- *Convenience and efficiency*: food for difficult situations – astronauts, the military, air passengers, refugee camps; quick and easy preparation of home meals and snacks.
- *Health/nutrition*: encouraging children to eat healthy foods; easy to prepare healthy snacks at home; food for people with dysphagia (chewing and swallowing difficulties) or other eating difficulties; food designed for individuals' personal nutritional needs; controlling food intake for weight management or special dietary needs (e.g. avoiding allergens or animal products).
- *Reducing waste and enhancing environmental sustainability*: more efficient storage, transportation and preparation of food; de-centralisation of food preparation; re-use of discarded food; generation of edible food packaging or cutlery.
- *Alleviating world hunger*: more efficient use of available foods; use of alternative food sources.
- *Ethical meat or meat substitute production*: using laboratory-cultured animal proteins or vegetable proteins to create meat substitutes that can be moulded by 3D printing technologies.

Most of the potential uses for 3D printed food are in development in technology companies, food industry or university research labs. Some 3D food printers are already in use in commercial settings, restaurants and nursing homes. 3D printed novelty sweets, biscuits and chocolates have been available to the public for several years (Lipton *et al*., 2015; Sun *et al*., 2015). An American restaurant chain has begun offering 3D printed pizzas fabricated by the Beehex printer (Hall, 2016), and several fine-dining restaurants have experimented with 3D printed dishes (Koenig, 2016). At the time of writing, no 3D food printers were available for domestic use. However, researchers and developers predict that these machines may soon be part of modern home kitchens, allowing people easily to prepare nutritious foods for themselves and their family members. For example, the Natural Machines company is developing its Foodini 3D printer for both restaurant and domestic use, advertising it as a 'new generation kitchen appliance' (Natural Machines, 2016).

In this chapter, we focus on two 'promissory narratives' (Stephens and Ruivenkamp, 2016) of 3D printed food: (1) reducing waste and enhancing environmental sustainability; and (2) alleviating world hunger. Both are endeavours that have been central to many recent food activist causes (Alkon and Norgaard, 2009; Evans *et al*., 2012; Rosin *et al*., 2013). 3D food printing is an area of technological innovation, therefore, that involves the digital technology industry and researchers taking up and engaging with food activism ideals and discourses as a way of envisaging and promoting these promissory narratives. From the first announcements of this nascent technology, industry sources, researchers and activists presented 3D food printing as capable of making a significant contribution to environmentally friendly, ethical and sustainable food habits (Lupton, in press).

In discussions of the possibilities of 3D food printing, the interests of the agri-food and digital technologies industries, researchers into foodways, agriculture and food science and food activists have converged. This convergence of interest has been evident in the development and promotion of novel food products and food production techniques in the past. Industry and entrepreneurs have seized opportunities to mobilise food activist concerns to generate interest in and support for their products (Rollin *et al*., 2011; Stephens and Ruivenkamp, 2016). For example, these issues were addressed by participants in a panel at the SXSW (South by South West) Eco Conference in Austin, Texas, in 2013, comprised of Jason Clay from the World Wildlife Fund, Hod Lipson, a 3D printing researcher at Cornell University and Jed Davis, Director of Sustainability at Cabot Creamery, an American dairy agricultural marketing cooperative.

As noted earlier, it has been speculated that 3D food printing technology has the capacity to reduce food waste by re-using materials that would otherwise be discarded or providing a mode of processing and preparing food that is more efficient. Some proponents have suggested that 3D food fabrication devices could be incorporated into the 'smart kitchen' (Davies, 2014), assisting people to reduce food wastage and maximise the nutritional value of the food they prepare (Lipson and Kurman, 2013). This view is evident in the words of Anjan

Contractor, whose company received a grant from NASA in 2013 to create a prototype of a 3D food printer. He was quoted in an online news story (Mims, 2013) as envisaging a future when food waste as part of food preparation could be eliminated by the domestic use of such a machine. Printed food also offers the promise of alleviating world hunger by using alternative food sources. The Dutch technology holding company TNO is at the forefront of research and development in 3D printing food, partnering with the food industry in working on innovations such as using alternative nutritional sources for food production, Researchers from TNO and elsewhere are working on 3D printing food techniques that use sustainable food sources such as insects, algae, duckweed, grass, lupine seeds and beet leaves to fabricate foods that are visually appealing to consumers and taste good (Mims, 2013; Soares and Forkes, 2014).

Novel food technologies and activism

The social, symbolic and emotional dimensions of our relationships with the substances that we ingest have significant implications for those actors and agencies who are attempting to encourage the adoption of novel food technologies into mainstream food supplies and consumption practices. Central to these efforts are the cultural meanings that shape decisions about what materials are considered 'food' and how these should be generated, processed, prepared and consumed. These meanings are entangled in multiple material, economic, cultural, ethical and affective domains.

Numerous studies have sought to understand consumers' views about unfamiliar foods and food preparation and processing techniques, and how and why these views form and manifest in practice in particular ways. As they have demonstrated, ideas about edibility and food choice are often unrelated to the digestibility or nutrition of substances. Humans tend to favour eating material that they have been acculturated to accept as 'food' within their social, cultural, geographical and historical contexts (Douglas, 1966; Rollin *et al.*, 2011; Stephens and Ruivenkamp, 2016). A vast and complex array of factors work to define the categories of 'food' and 'non-food'. Certain substances are considered more appropriate for either gender to consume as food. Food definitions are also categorised by age, religious or ethical beliefs and ideas about a 'healthy diet' as well as customs concerning what food should be eaten at certain times of time and which types of plants and animals are considered edible and which are not (Ashley *et al.*, 2004; Falk, 1994; Lupton, 1996; Messer, 2007; Mintz and Du Bois, 2002; Shelomi, 2015).

The introduction of radical new ways of processing and preparing foodstuffs evokes a range of sociocultural and emotional responses. This is evident in the discourses and practices employed by food activist groups and organisations as well as among consumers in general (Bredahl, 2001; Marcu *et al.*, 2015; Messer, 2007; Rollin *et al.*, 2011; Stephens and Ruivenkamp, 2016; Tenbült *et al.*, 2005). In an age in which high technological processing is typically portrayed among nutritionists, gourmands, public health professionals and many food activist groups as producing unhealthy, less nutritious and tasteless food, the combination of new technologies with food tends to be viewed with suspicion. The products of novel food technologies are commonly represented as a violation of the natural order of things that is likely to lead to a dystopian future. What Rozin and his colleagues (Rozin, 2005; Rozin *et al.*, 2012) characterise as 'the natural preference' – the desire to consume foods that are considered to be 'natural' in their production and processing, involving little perceived human interference and few additives – is dominant in most responses to novel food technologies.

The associations of novel food technologies with the unnatural persist in the face of contentions that these technologies have the potential to reduce negative environmental impacts and facilitate more efficient and sustainable food production and distribution (and therefore to be supportive or protective of the 'natural' world) (Azadi and Ho, 2010). Such a position is evident, for example, in campaigns against genetically modified (GM) foods spearheaded by groups such as Greenpeace, which have achieved particular success in Europe. In these campaigns, activists sought to represent GM foods as 'non-foods' due to the scientific manipulation of their genetic make-up involving the interspecies transfer of genes (Hellsten, 2003; Messer, 2007). The term 'Frankenfood' has been used in activists' attempts to challenge the introduction of GM foods, as well as such practices as bovine growth hormone used to increase the production of milk in cows (creating 'Frankenmilk') (Hellsten, 2003) and the production of cultured meat in the laboratory (Marcu *et al.*, 2015; Stephens and Ruivenkamp, 2016). In its direct reference to the well-known disturbing figure of Frankenstein's monster, this term draws attention to the departure from the natural represented by these foods and also evokes the monstrous distortion of nature by science.

These types of portrayals of novel food technologies in well-established food activism discourses in some ways work against their attempts to encourage consumers to accept the promissory narratives of 3D printed food. Yet, despite these often entrenched notions about 'food' and 'non-food', histories of food consumption clearly demonstrate that changes in food preferences and choices are possible within specific sociocultural contexts (Burnett, 2013; Lupton, 1996; Mintz and Du Bois, 2002). Ideas about edibility are open to contestation and transformation. While humans tend to be conservative in their food definitions and choices, many also seek novelty and excitement. This is described as the tension between neophilia and neophobia; or, as Fischler (1988) puts it, 'the omnivore's paradox'. Novel foods are often associated with excitement, high status and adding value to one's life (Lupton, 1996). The willingness to try new foods can denote sophistication, adventurousness and evidence of being a 'true gourmand' (Falk, 1994), and is often a selling point in tourism marketing to foreign countries where the cuisine may be unfamiliar to visitors.

Beyond these inducements to try new foods, health, moral, ethical or political concerns can impel changes in food habits and preferences (Messer, 2007). People can be willing to try new foods if they can see benefits for themselves, such as lower cost or better taste or health qualities, or for addressing global concerns such as environmental sustainability and food shortages (Rollin *et al.*, 2011). Developers and researchers have promoted novel food production methods such as cultured meat using these kinds of narratives (Stephens and Ruivenkamp, 2016). Food activism can play a role in effecting this change, albeit in concert with a host of other actors and agencies, such as the medical profession, government and public health authorities, advertising and marketing strategies and the news media. To do so, however, they first need to be aware of what members of the public are currently making of this technology and how receptive they may be to the idea of consuming it or serving it to family members or friends. Our study was designed specifically to address these topics.

Details of the study

An online focus group discussion was undertaken with 30 Australian participants over a four-day period in March 2016. We chose to use this method as a way of readily accessing participants from a diverse range of backgrounds and living in different areas of Australia. The project was approved by the human ethics research committee of the University of Canberra.

A research company specialising in online research hosted the discussion on their customised platform. The participants were recruited from panels of people who had previously signed up to be involved in research projects hosted by the company and other associated market research companies. Panel members were sent an invitation by the market research company to participate in the study. The resultant group included 19 women and 11 men, with 13 participants aged between 18 and 39, and 17 aged 40 and over. There was a mix of household types, including single people (9 participants), those who were part of couples with no children living at home (7) and those with children aged under the age of 18 living at home (12), while two participants had adult children still living at home. The participants' occupations were diverse, including students (2), tradespeople or labourers (2), professionals (9), clerical/service workers (4), managers/administrators (4), stay-at-home parents (3), business owner (1), carer (1) and retired people (4). They came from all seven states of Australia.

People who agreed to participate were asked to read an online information and consent form. Once they had ticked the box notifying that they wanted to participate, they were provided with a link to the study platform. All participants were requested to respond to the questions by typing in their answers in the spaces provided on the online page. They were able to view other participants' answers and comment on these as well. No real names were used in the forum, and we chose pseudonyms for all participants. Once they had completed all the questions, including responding to other participants' comments if they wanted, the participants were compensated for their time with a gift voucher worth AU$50.

The discussion questions began by asking participants about their general food preferences and choices, what they knew about 3D printing in general and more specifically about 3D food printing. This was followed by a series of questions about different types of 3D printed food, including a section where participants were asked to respond to images of seven different printed food stuffs that have already been fabricated by various entrepreneurs or designers. We provided the research company with the questions and images used and conducted all analysis of the research material, which included the typed-in responses to our questions. This material raised many interesting issues concerning the participants' attitudes to 3D printed food. We began by grouping responses under topics. We then identified and analysed the discourses and themes underpinning the respondents' rationales for their responses, many of which were shared across the topics.

Factors underpinning general food choices

We began the focus group discussion with general questions about the current food choices that the participants make as a means of contextualising their later responses to 3D printed food. The first question prompted participants to comment on their motivations for current food choices. They were asked to identify the five most important factors when making these choices and to explain why these were important to them. They were then asked to discuss how easy they found it to buy and prepare the food they preferred to eat. This was followed by questions about how much their food choice was determined by what others in their family would eat and by other factors (locally sourced food, allergies or intolerances, ethics, religion or anything else).

The responses revealed that the factors of the taste of food, its quality, healthiness/nutritious qualities and freshness and its price or value for money were the most important when participants were making their current health choices. Nearly everyone said that they had no difficulties in finding and preparing their favoured foods. Some people's food choices were influenced by religious beliefs (e.g. prohibitions on eating beef and pork), by needing to cater to small children's or partners' food preferences and by cultural preferences (one participant with an Asian background gave the example of his family eating a lot of rice).

Issues concerning environmental sustainability were not foremost in the participants' identification of what they sought in the food they ate. David was the only participant to specifically mention environmental issues when identifying the factors that shaped his food choices, expressing his desire to source 'environmentally friendly' food. Four others mentioned seasonal eating as a priority, as they were keen to consume food that was both fresh and less expensive. Five participants said they preferred organic food, largely for its perceived attribute of being healthier because of less contamination by such chemicals as pesticides and antibiotics.

Only two participants – Anna and Sarah – self-identified as vegetarian. Anna also singled out seasonal and organic attributes of food as considerations in her food choices, as well as her desire to consume 'ethically produced and fair-trade' food. Again, however, her decisions to do this were dependent on cost and her available budget. These environmental and ethical concerns were expressed by a limited number of respondents, but this preference was not the deciding factor in food purchasing behaviours. As the other vegetarian, Sarah stated: 'I definitely prefer to buy things that are responsibly sourced and that are ethically grown, etc., but I don't limit my purchases in an attempt to only buy sustainable items.'

General knowledge of and responses to 3D printed food

The next set of questions asked the participants what they knew about 3D printing technologies generally, and then about 3D food printing more specifically. It was clear from their responses that these technologies were unfamiliar to the participants. While some of the participants had heard of 3D printing technologies, they had few inklings of what it might involve. Comments such as the following were common: 'I have not heard of a 3D printer but am quite inquisitive as to what it actually does' (Rajah). Only one person, Kylie, had seen an actual 3D printer (used by her spouse for hobby activities), but some remembered media coverage of the technology: 'I don't know a lot about 3D printing. But I have seen 3D printers on TV printing everything from toys to simple medical parts for surgery' (Anna).

The participants were even less sure of what 3D printed food might involve. Some engaged in some wild speculation trying to imagine how this process might work, as in the examples below:

> Never heard of food made by 3D printers and I'm a little shocked!!! How can this even be real??? This is outside of my expertise, no idea how this can work! A chef hiding in a box maybe?
>
> (Sally)

> Goodness this gets even more interesting. 3D printing for food – how does this work? Does the printer cook food for us or prepare food for us?
>
> (Rajah)

> I have not heard of 3D printers in food industry and I am not sure how they are gonna use it. Not sure how they will merge it in food industry, but it will be interesting to see.
>
> (Hanu)

Several people assumed that the food would be somehow 'plastic' and therefore inedible. This assumption was clear in the words of Kylie, who said: 'I haven't heard of any food being made with a 3D printer. The printer uses plastic to "print", therefore the food would be inedible.' This lack of knowledge about how food might be printed was often coupled with uncertainty about how such foods could be defined or identified. Bill, for example, noted that he knew nothing about 3D printed food and was unsure about whether using this technology to process food was appropriate: 'I haven't heard of any examples of food made by 3D printers. I have no idea on how it might work and don't think food is an apt use of the technology.' Even when Bill was introduced to some examples of what 3D printed food might be (by virtue of the images of such food we provided), he made it very clear that 'it was not the sort of food' that he would eat. When asked about the possible good and bad aspects of the technology, he responded only in the negative. Bill was concerned about 'contamination from the plastic or printer itself, the health and safety risks … poor taste, artificial, not safe'.

Re-using waste

We noted earlier that environmental sustainability issues were not very important to the participants when making choices about the food they consumed. It is not surprising then, that the potential environmental benefits of 3D printed food were rarely mentioned in the focus group discussion, appearing overtly only three times. One participant who referred to these benefits had already expressed pro-environmental beliefs and practices, while two others nominated environmental credentials as a positive for the printed food, especially its capacity to process alternative ingredients into food. Of the other five people who expressed a concern for environmentally sustainable foods, all but one (Bill) were willing to consider consuming 3D printed food as long as its healthiness and safety were assured.

The broader problem of food waste was acknowledged by several participants when we asked them directly if they thought 3D printing could offer a viable solution. However, they raised concerns about the potential costs of using technologies to produce 3D printed foods when householders could take other action to reduce waste without the need for special technologies. In these accounts, unfamiliarity with the 3D printing process underpinned reservations about the safety of using foods that would otherwise be discarded as waste. Wider questions of how such food would be collected, stored and processed to ensure food safety were foremost when people explained their answers. As Sarah noted:

> Hmm, I am very committed to preventing waste, but I don't think I could try food that could have some risks associated with it, regarding bacteria and other germs. It's a fantastic idea in theory though! I think a lot of trial and error would have to be done before introducing this particular step to the population.

Other people displayed more extreme reactions to the idea of re-using food waste in printed food products. Bill was very wary of the need to process food waste to render it more edible: 'It sounds disgusting. Why would there be a need to make something more edible? That sounds even more disgusting. If it is left-over, it is not fresh and may not be even safe to eat.' Even though Bill did express environmental concerns about the food he ate earlier in the focus group, the perceived artificiality of the 3D food printing process and the possible health risks he envisaged prevented any identification of possible environmental benefits for this type of food.

Using alternative food sources

Enhancing environmental sustainability and alleviating world hunger by using alternative food sources raised another set of issues. Very few participants expressed positive responses to the idea of consuming printed foods made from algae or insects. Here again, they made little explicit mention of the potential environmental benefits such foods could have. Two of the more positive reactions came from Emma and David. Emma identified that consumption of algae and insects 'could also prove to be a sustainably sourced and nutritious ingredient made more appealing by this process'. David observed: 'With current population growth and the need for new food sources this could be the way to go.' However, both qualified their statements. Emma was concerned that choice about whether or not to consume these foods should be enabled through transparency (such as food labelling protocols), while David suggested that there may allergen risks in food made from such ingredients.

The more typical response from participants to the idea of food made from these alternative ingredients was outright disgust. Indeed, many used this word to describe how they felt about eating printed foods made from algae and especially insects. This response was most evident in the section of the discussion where participants were asked to give their responses to an image we provided of a crunchy savoury snack food that had been fabricated from ground insects. When they were asked whether they would eat this food themselves or serve it to family members or friends, most participants were vehement that they would avoid either eating it or serving it. Comments such as the following are examples of such reactions:

> I think it's very natural and probably healthy for you as it's ground insects, but I'm not sure if I would like the taste, I have never eaten insects before.
>
> (Marie)

> I saw the word insect and was instantly disgusted by it. Even if it was considered natural and edible by others, I would wonder how it was put together. Very strange.
>
> (Sally)

As these excerpts suggest, because the snack food was made of insects, it can be considered 'natural', but the unfamiliarity of this ingredient, together with what was considered to be an unnatural processing method, rendered it 'very strange', 'disgusting', and therefore a non-edible substance. Insects are not usually considered to be 'food' in Australia and other Global North food cultures, regardless of how they are prepared (Shelomi, 2015), and the very idea of eating them proved challenging to nearly all our participants.

The appearance of printed food

It was evident that the appearance of the insect snack food was almost as off-putting as its ingredients. Even though none of our questions directly asked for participants' comments on the appearance of printed food, this was an important dimension to which participants had strong reactions. The insect snack food had been fabricated in geometric shape and was off-white in colour. It thus looked unfamiliar and did not present overtly as a food stuff. Comments such as: 'This looks like some concrete structure or the inside of the drain part in a washing machine – extremely off-putting' and 'This snack looks very artificial, unhealthy and disgusting' were articulated.

It was not only the insect snack that aroused disgust, however. The jellied appearance of some of the other foods we showed images of to the participants (e.g. a meal made with pureed chicken and vegetables) was off-putting to quite a few of them. They found the jelly-like texture to be unappealing and also unnatural, as it was thought that meat and vegetables should not have this kind of texture. This food was described as looking 'strange', 'too processed', 'slimy' or 'too soft'. Here again, such repugnance to slimy-textured food is a dominant norm in the Global North (Lupton, 1996; Martins and Pliner, 2006). Several people also commented unfavourably on the brightly coloured confections they were shown, remarking how artificial and plastic they looked, and therefore not like 'real food'.

Repugnance at the appearance of such printed food emerged elsewhere in the discussions as well. For example, Emma made clear her preference for 'natural things'. For her, all printed food, regardless of its ingredients, was categorised as 'artificial'. She noted that 'it all sounds very synthetic' and it was the appearance rather than the content of printed food that she found challenging to accept. For Emma, the notion of something artificial masquerading as the 'real thing' was particularly disturbing:

> [t]he bit about trying to make a non-natural food look like a natural food freaks me out a bit too. I think I would rather it not try to look like something from an actual animal and not pretend to be something it's not.

This attitude seems to be fuelled by concerns that consumers could be duped into ingesting something that mimics the natural product. This process of artificial replication means consumers would be less likely to know (or, more specifically, less inclined to question) what processes had been used to produce the food and, thus, might be more likely to consume it. According to this line of reasoning, that which was once familiar becomes strange, throwing into doubt the extent to which consumers can exercise choice and control over their diets. As Anna observed: 'My comfort level rose the more the products resembled actual whole food and recognisable meals, which was disturbing. It is easier to feel suspicious about something that looks foreign and alien.' Indeed, in response to this concern about being able to identify the constituents of printed food, one participant suggested that insect-based products should be manufactured in the form of a large, intact insect, thereby overtly signalling its provenance.

Overall benefits and drawbacks

The participants were asked to state what they saw as being the possibly beneficial aspects of 3D printed foods as well as their drawbacks. Overall, most participants expressed scepticism about the use or value of 3D food printing technologies to their own lives. There was little evidence that people were willing to introduce these foods into their daily diets. Several participants were willing to consider consuming printed food once they had been exposed to some of the examples we provided. However, here again the nutritious qualities of the food were identified as very important. Participants reiterated their opinions that the safety of the production process would have to be ensured in terms of possible contamination of food by chemicals or bacteria, the extent to which the foods contained such additives as preservatives or whether nutrients would be lost during processing by the technology.

For many people, the extent of processing that appeared to be involved in printed food was a concern. As Steve commented:

> It's probably pretty expensive, and there might be extra artificial ingredients and additives in there that may not be the healthiest … [It] can't possibly be as healthy. It would either be the same, or worse. Like how can you beat a carrot with a printed carrot as far as nutrition goes? Impossible. Artificial ingredients and additives used that you don't need in already natural products.

Most of those who were open to consuming printed food saw themselves as adventurous eaters, willing to try anything once. For others who may have been attracted to trying something new, the processed nature of the food was unsettling enough to make them feel ambivalent. As Anna noted:

> The pictures and the topic I find both fascinating and disturbing. I prefer my food to come from natural whole ingredients by a regular growing process, versus technological manufacture, and I feel that something inherently valuable is surely lost in this process.

One of the questions asked the participants whether they thought 3D printed food might benefit certain social groups. They were provided with a range of specific groups (pregnant women, the elderly or those who are sick, young children, athletes in training or others) as prompts. The most common group identified as being able to benefit from was the elderly who experience difficulties swallowing and lack variety in their foods. While we did not suggest the socio-economically disadvantaged in our list, several participants spontaneously identified 'the poor', 'the starving' or 'the homeless', both locally and globally, as groups that could benefit from access to printed food.

It was clear, however, that while printed food may be suitable for these others, most participants were unwilling to consume the foods themselves or to offer it to their family or friends. This was exemplified by Marie's statement: 'I would not like to eat 3D food at all. The idea is not appealing to me. It might be good for the homeless.' Later she further noted that '[i]t might be good for the really poor as it might in time be a cheap way to make food' and '[i]t would probably need less material to make a meal'. Similarly, while Kirsten could not see a benefit for herself or her family, she viewed 3D food printing as possibly helping people living in countries where food was scarce: 'Not sure if there is a good side to it apart from being able to help countries that can't grow enough food.' Emma similarly observed that there 'may be a place for this if used to feed those who can't otherwise access [food]'. 3D printed food technology, therefore, was seen to be applicable primarily in situations where people had limited control and choice of their food supplies. For the underprivileged or the ill, it was seen as an opportunity to expand their choice, while for the participants themselves, it was commonly seen as a threat to their personal choice and control over food.

Discussion

Our research demonstrated that 3D printed food introduces a number of unfamiliar elements to the production of potentially edible substances. First is the way in which these substances are processed and prepared. 3D food fabrication uses a new digital technology of which few people have yet heard, let alone have direct experience. Second is the apparent extent of processing that substances undergo as part of the fabrication process. Given that the process is not well understood by members of the public, they are unsure of the nature of this processing and its possible benefits or ill-effects. Third, the appearance of the some of the products generated by this technology are considered unusual, strange or having an unacceptable texture. While some printed foods mimic the appearance of familiar foods or are clearly identifiable as being composed from these foods, they are also often overtly different. A fourth element of unfamiliarity is the content that is used to generate these substances. Many printed foods are produced using familiar and long-accepted ingredients such as sugar, chocolate, dough and food purees. Others, however, are generated from ingredients that are traditionally considered non-edible and unacceptable as foods in western cultures – insects, algae, duckweed, food waste, and so on.

Our study findings demonstrate that all of these elements contributed to people's wariness about accepting 3D printed food as safe and edible. When several of these elements are present, as in the case of printed food products made from insects, participants' emotional responses were most vehement. Other factors that contributed to negative responses from participants drew on contemporary preoccupations about the healthiness of food in general: how much sugar, salt, fat or preservatives it contains, how safe the conditions are in which it is manufactured; what allergens or bacteria or other contaminants may be present; and how the processing of the ingredients may destroy nutrients.

Overall, environmental issues played a minor role both in people's current food provisioning practices and their identification of the potential benefits of

3D food fabrication technology. While most consumers in this study expressed a preference for food that was 'natural', efforts to reduce their own impacts on the 'natural' environment were extremely limited. However, we did find that when they were presented with information outlining potential environmental benefits of printed food, more people expressed an environmental ethic and willingness to try this product. Here again, this response was primarily related to foods that were already familiar, rather than new food sources such as insects, and people were more likely to view the foods as beneficial for disadvantaged others than for themselves.

On the one hand, responses to 3D printed food evoke the desire for food and the mode of its production or preparation to be familiar. This type of food production can be perceived as highly unnatural because of its strangeness and its association with high technology and extreme levels of manipulation of foodstuffs. People tend to be more suspicious of food production technologies that are least familiar to them. This has been demonstrated in previous research on attitudes to nanotechnology, irradiation of food, animal cloning and GMO foods (Rollin *et al.*, 2011) and laboratory cultured meat (Marcu *et al.*, 2015; Stephens and Ruivenkamp, 2016). On the other hand, 3D printed food offers the promissory narratives of novelty and potential to manufacture food with less waste, more tailored nutrients and edible matter that is ethically acceptable. Such foodstuffs thus could be perceived as promoting the goodness of nature (as in improving environmental sustainability or human health and well-being, or alleviating world hunger). These are the narratives that could be employed by food activist attempts to introduce this technology to the public. Yet these attributes were not recognised by many of the participants, for whom the idea of 3D printed food was foreign and unsettling, even disgusting.

Not only was the technology used viewed as a very artificial and highly processed way of preparing food, participants also found several of the printed foods we showed them to be unacceptably 'unnatural'-looking. This was because of their texture (too slimy) or conversely, because they appeared too brightly coloured, plastic or hard-looking, or too 'perfect', as if made from non-edible materials. Even when the printed foods look reasonably similar to familiar foods, this can be viewed with suspicion because they are 'uncanny': both too like and yet too different from the familiar (Schoenherr and Burleigh, 2015).

The potential to use fresh ingredients to prepare printed foods was not considered by our participants, many of whom valued this quality in the food they chose to eat. Even though we notified the participants of what each food we showed them was made from, the highly manipulated, unusual and artificial appearance of the foods tended to belie the fact that some of them were made from fresh ingredients that had simply been cooked and pureed or otherwise liquified. So too, the use of unfamiliar ingredients such as insects (which most people would agree are 'natural' entities) aroused disquiet. In both cases, this technology involves 'matter out of place', as Douglas's (1966) classic taxonomy of purity and contamination has it. In the words of Anna, they were 'both fascinating and disturbing'.

Conclusion

3D printed food offers some possibilities to facilitate food activist efforts directed at improving environmental sustainability and nutrition for all social groups and the alleviation of world hunger. However, many of these possibilities remain speculative, and the finer details of how they may come to be introduced as novel food technologies remain unclear. Thus far, techno-utopian promissory narratives about 3D food printing have dominated in popular media, industry outlets and academic forums. The opinions of food activists or any others who may disagree with introducing 3D food printing into general use have received little coverage (Lupton, in press).

In identifying how our participants responded to the idea of consuming 3D printed food or serving it to others, our findings go some way to introducing a more nuanced perspective. They suggest that efforts to promote use of 3D food printing technologies, whether by activists or other actors and agencies, must engage with the persistence of a consumer preference for 'natural foods' and the importance of their immediate affective and visceral reactions to food, including such elements as what kind of technology is used to process it, what the food is perceived to be made of, how fresh it is, the degree of processing it undergoes and how many other elements may be added to create it. As we noted earlier, the preference for the natural is itself a key plank of many food activist efforts themselves, and activists must therefore come to terms with what may be considered to be significant contradictions in promoting a novel food technology like 3D printing that currently is viewed in many respects as 'unnatural'.

The entrée of digital fabrication technologies into foodways, as we have demonstrated, is transgressive and confronting. Many culturally entrenched notions of what food is, what it should be made of, how it should look and how it should be processed or prepared are challenged by this novel technology. Our research findings suggest that to maximise the potential benefits of 3D food printing for activism, there is a need to shift the focus of the technology beyond niche groups (such as the elderly, the poor and those living in areas where food supplies are limited) and to emphasise its qualities for better nutrition and the positive dimensions of its novelty. The maturity of the technology may well lead to better familiarity and thence to greater acceptability of printed food.

In the meantime, the public's suspicion of the process and the products – and their frank disgust at some of the possibilities of 3D printed food – may be alleviated if food activists who want to promote this type of food processing can work to allay people's concerns by explaining in detail how it works, what ingredients are used and how food safety measures can be ensured. In addition to these qualities, however, we suggest that food activists could emphasise how other benefits may be offered by this technology: for example, the ways in which it can contribute to innovative gourmand meals, make meals look more attractive, offer acceptable new flavours and enhance the freshness and nutritious qualities of foods. If the quality, nutritious properties and freshness of the ingredients can be guaranteed and rendered transparent to consumers and the appearance of the food manipulated so that it is acceptable to them, it is likely that people will accept 3D printed foods more readily.

References

Alkon, A. H. and Norgaard, K. M. (2009) Breaking the food chains: An investigation of food justice activism. *Sociological Inquiry*, 79: 289–305.

Ashley, B., Hollows, J., Jones, S. and Taylor, B. (2004) *Food and Cultural Studies*. London: Routledge.

Azadi, H. and HO, P. (2010) Genetically modified and organic crops in developing countries: A review of options for food security. *Biotechnology Advances*, 28: 160–168.

Bredahl, L. (2001) Determinants of consumer attitudes and purchase intentions with regard to genetically modified food: Results of a cross-national survey. *Journal of Consumer Policy*, 24: 23–61.

Burnett, J. (2013) *Plenty and Want: A Social History of Food in England from 1815 to the Present Day*. London: Routledge.

Davies, A. R. (2014) Co-creating sustainable eating futures: Technology, ICT and citizen–consumer ambivalence. *Futures*, 62(Part B): 181–193.

Douglas, M. (1966) *Purity and Danger: An Analysis of Concepts of Pollution and Taboo*. London: Routledge & Kegan Paul.

Evans, D., Campbell, H. and Murcott, A. (2012) A brief pre-history of food waste and the social sciences. *The Sociological Review*, 60: 5–26.

Falk, P. (1994) *The Consuming Body*. London: Sage.

Fischler, C. (1988) Food, self and identity. *Social Science Information*, 27: 275–292.

Hall, N. (2016) Try 3D printed food in a restaurant today! *3D Printing Industry*. Available at: http://3dprintingindustry.com/news/pizza-legend-turns-3d-printer-84590/ (accessed 16 August 2016).

Hellsten, I. (2003) Focus on metaphors: The case of 'Frankenfood' on the Web. *Journal of Computer-Mediated Communication*, 8. Available at: http://onlinelibrary.wiley.com/doi/10.1111/j.1083-6101.2003.tb00218.x/full (accessed 15 June 2016).

Koenig, N. (2016) How 3D printing is shaking up high end dining. *BBC News*. Available at: www.bbc.com/news/business-35631265 (accessed 16 August 2016).

Lipson, H. and Kurman, M. (2013) *Fabricated: The New World of 3D Printing*, Indianapolis, IN: John Wiley & Sons, Inc.

Lipton, J. I., Cutler, M., Nigl, F., Cohen, D. and Lipson, H. (2015) Additive manufacturing for the food industry. *Trends in Food Science & Technology*, 43: 114–123.

Lupton, D. (1996) *Food, the Body and the Self*. London: Sage.

Lupton, D. (in press) 'Download to delicious': Promissory narratives in coverage of 3D printed food in online news sources. *Futures*.

Marcu, A., Gaspar, R., Rutsaert, P., Seibt, B., Fletcher, D., Verbeke, W. and Barnett, J. (2015) Analogies, metaphors, and wondering about the future: Lay sense-making around synthetic meat. *Public Understanding of Science*, 24: 547–562.

Martins, Y. and Pliner, P. (2006) 'Ugh! That's disgusting!': Identification of the characteristics of foods underlying rejections based on disgust. *Appetite*, 46: 75–85.

Messer, E. (2007) Food definitions and boundaries. In J. Macclancy, J. Henry and H. Macbeth (eds) *Consuming the Inedible: Neglected Dimensions of Food Choice*. New York: Berghan Books.

Mims, C. (2013) The audacious plan to end world hunger with 3-D printed food. *Quartz*. Available at: http://qz.com/86685/the-audacious-plan-to-end-hunger-with-3-d-printed-food/ (accessed 9 April 2016).

Mintz, S. W. and Du Bois, C. M. (2002) The anthropology of food and eating. *Annual Review of Anthropology*, 31: 99–119.

Mohr, S. and Khan, O. (2015) 3D printing and its disruptive impacts on supply chains of the future. *Technology Innovation Management Review*, 5: 20–25.

Natural Machines (2016) *Natural Machines/FAQs*. Available at: www.naturalmachines.com/faq/ (accessed 14 June 2016).

Petrick, I. J. and Simpson, T. W. (2013) 3D printing disrupts manufacturing. *Research Technology Management*, 56: 12.

Rollin, F., Kennedy, J. and Wills, J. (2011) Consumers and new food technologies. *Trends in Food Science & Technology*, 22: 99–111.

Rosin, C., Stock, P. and Campbell, H. (eds) (2013) *Food Systems Failure: The Global Food Crisis and the Future of Agriculture*. London: Routledge.

Rozin, P. (2005) The meaning of 'natural'. *Psychological Science*, 16: 652–658.

Rozin, P., Fischler, C. and Shields-Argelès, C. (2012) European and American perspectives on the meaning of natural. *Appetite*, 59: 448–455.

Schoenherr, J. R. and Burleigh, T. J. (2015) Uncanny sociocultural categories. *Frontiers in Psychology*, 5. Available at: http://journal.frontiersin.org/article/10.3389/fpsyg.2014.01456/full (accessed 27 February 2017).

Shelomi, M. (2015) Why we still don't eat insects: Assessing entomophagy promotion through a diffusion of innovations framework. *Trends in Food Science & Technology*, 45: 311–318.

Soares, S. and Forkes, A. (2014) Insects au gratin: An investigation into the experiences of developing a 3D printer that uses insect protein based flour as a building medium for the production of sustainable food. In E. Bohemia, A. Eger, W. Eggink, A. Kovacevic, B. Parkinson and W. Wits (eds) *Proceedings of the 16th International Conference on Engineering and Product Design*, 2014 University of Twente, the Netherlands. pp. 426–431.

Stephens, N. and Ruivenkamp, M. (2016) Promise and ontological ambiguity in the in vitro meat imagescape: From laboratory myotubes to the cultured burger. *Science as Culture*, online first.

Sun, J., Zhou, W., Huang, D., Fuh, J. Y. and Hong, G. S. (2015) An overview of 3D printing technologies for food fabrication. *Food and Bioprocess Technology*, 8: 1605–1615.

Tenbült, P., De Vries, N. K., Dreezens, E. and Martijn, C. (2005) Perceived naturalness and acceptance of genetically modified food. *Appetite*, 45: 47–50.

9 Hashtag activism and the right to food in Australia

Alana Mann

Introduction

> Bird chirps sound meaningless to us, but meaning is applied by other birds. The same is true of Twitter; a lot of messages can be seen as completely useless and meaningless, but it's entirely dependent on the recipient.
>
> (Twitter co-founder Jack Dorsey, cited in Sarno, 2009, paragraph 22)

Social media tools enable today's activist to engage in fluid collaborations, integrative campaigning and the development of alternative media platforms that represent 'new spaces of social movement activism' (Olesen, 2005: 419). The Internet-based issue-networks that grow in these spaces play a vital role in 'disclosing such an assemblage of actors jointly implicated in an issue that no agent, no organisation is effectively taking care of' (Marres and Rogers, 2005: 929). Food security is one of these issues in Australia, where up to 25 per cent of households in disadvantaged communities are affected (Right to Food Coalition, 2016).

The practices of food security advocates are increasingly shaped through the affordances of digital media platforms. Tools that allow users to cluster, share and modify messages facilitate the networking of actors engaged in campaigning. Hashtag activism, coined by *New York Times* journalist David Carr (2012), allows users of Twitter to focus tweets on a single issue. Prominent examples include large-scale mobilisations around the hashtags #arabspring, #occupy, #Idlenomore and #BlackLivesMatter. This study applies framing theory to better understand how one particular group of food security advocates in Australia employed Twitter in a campaign designed to have an impact on policy regarding the right to food. Explaining how members of the Right to Food Coalition, a 'national food security workforce' (Right to Food Coalition, 2016), attempted to drive collective, and connective, action on food insecurity via the micro-blogging social networking platform, this research reveals how activist practices are exploiting digital media affordances.

Using the Twitter Capture and Analysis Tool Set (TCAT), I conducted a frame-critical analysis of Twitter coverage of a Right to Food Coalition campaign in Australia over a three-day period in June 2016, leading up to the federal election. This timing was significant as the campaigners aimed to raise awareness about household food insecurity and lobby politicians to drive policy that ensures access to affordable and healthy food for all Australians. The study provides insight into how Twitter elements such as hashtags, handles and mentions can serve as issue-framing devices and mediators within food advocacy networks. It also demonstrates the strategic value of the tool for facilitating a cross-flow of information between ideologically aligned advocacy organisations, both domestically and internationally. These affordances contrast with Twitter's limitations in facilitating political conversation and participation.

Social media and activism

Media impact on society and culture through their scale, speed and character of interaction (Borra and Rieder, 2014). The study of social media, now so deeply embedded in our media ecology, is part of a broader agenda to study society *through* the Internet (Rogers, 2009). Recognised as phatic, i.e. serving a social function (Miller, 2008), our social media practices provide a way for us to connect and relate to one another. The affective and ambient modes of expression we use on platforms like YouTube, Twitter, Yelp, Instagram, Pinterest and Facebook can produce feelings of community, belonging and solidarity (Papacharissi, 2015). The explosion of virtual food and 'foodie' (Rousseau, 2012) communities is evidence of how these platforms promote the sharing of our everyday experiences of eating; as Signe Rousseau says, 'social media do what food does best: they bring people together' (2012: 5). In virtual food communities, political activism coincides with recipe sharing and restaurant reviewing while also facilitating local, place-based mobilisations. Platforms such as Twitter contribute to the growing sphere of global social media activism (Cottle and Lester, 2011) by providing new opportunities for the articulation of concerns and dissent through transnational advocacy networks (Keck and Sikkink, 1998). The 'polyvocal protest' (Ruiz, 2014) of activists includes campaigns for Fair Trade and food sovereignty; public health communication to prevent lifestyle-related diseases, like diabetes; new food foraging apps that point consumers to ethical and local produce; and food hackers seeking to liberate new food technologies for the greater good.

In these examples new communication technologies enable actors other than elites to 'subvert the practice of communication by occupying the medium and creating the message' (Castells, 2012: 9). The Internet enables ordinary people to challenge gatekeeper functions and become 'media producers' (Couldry, 2000). Activists can avoid frame 'traps' or misframings that lead to ambiguity, error and misinterpretation (Goffman, 1974). Additionally, conventional media frequently pick up feeds from independent media platforms and social media sources such as Twitter and Facebook, as witnessed in the cases of Put People First (PPF), the Arab Spring, Spain's 15M, the *indignados* and the Occupy Movements (Bennett and Segerberg, 2012).

The relationship between social media and civic engagement, however, is not always clear-cut. The persistence of digital divides, the prevalence of elite voices and the commercialisation of the Internet raise questions about the democratising potential of social media. Sustained political and social engagement is challenged by notions such as slacktivism and clicktivism where sharing, joining, liking and even donating through social media represent illusionary, ephemeral and unsustainable participation that limits deeper analysis of issues and real engagement in political organising (Morozov, 2009). Further, Web 2.0[1] has led to the fragmentation of those 'formerly known as the audience' (Rosen, 2006) into interest groups based on online profiles and preferences. This challenges the notion of a public sphere where all citizens engage in critical public debate over matters of general interest (Habermas, 1989).

Public spheres are places and spaces where private individuals can share their opinions and come to consensus, or not, about critical issues. The efficacy of civil society in monitoring the state's power lies in its ability to take new social issues from the periphery of the public sphere and move them to the political centre. In this idealised democratic public sphere, consensus is reached through deliberative decision-making. The notion of the contemporary, democratic public sphere as 'an intermediary system of communication between formally organised and informal face-to-face deliberations in arenas both at the top and at the bottom of the political system' (Habermas, 2006: 9) emphasises the participation of ordinary citizens in political discourse on issues of social importance. The degree of atomisation or fragmentation of the public sphere, and access to it, directly influence opportunities for claims-making and the effectiveness of 'contentious politics' (Guidry and Sawyer, 2003) in debates on issues like food security. Given these constraints, virtual communities of activists can be characterised as digital echo chambers, or pseudo public spheres that create and support a 'homogenous climate of opinion' (Fuchs, 2014: 199). If the limitations of social media to facilitate wider conversations within a democratic public sphere preclude the possibilities for meaningful political engagement, what other possibilities do the affordances of platforms such as Twitter offer digital food activists?

Twitter as hashtag activism

Twitter is one of the top ten global social networking sites with 316 million monthly active users generating 500 million tweets per day (Socialbakers, 2016). It has evolved from an 'awareness system' (Kwak *et al.*, 2011) to a platform for public participation and debate on specific societal issues. The study of 'Twitter II' as an object seeks to 'de-banalise' (Weller, *et al.*, 2014) the platform by highlighting its value as a way to study cultural conditions and trace the evolution of an issue through 'professional communities of practice' (Turoff and Hiltz, 2009). This epistemological shift reflects the evolution of the platform; notably a development from inconsequential, friendly tweets to event following, advocacy and mobilisation.

Changing perceptions are marked by increasing attention to Twitter from mass media and communication researchers as a platform for political discourse and activism (Maireder and Asserhofer, 2014). Media scholars claim that to argue Twitter's 140-character limit per tweet renders it 'almost completely devoid of substantive content' (Miller, 2008) is to deny its value as news media. Users broadcast or narrowcast to followers (Kwak *et al.*, 2011). News consumption takes on a new 'social dimension' as 'by sharing and commenting on news stories, ordinary citizens frame and potentially change news agendas' (Jensen *et al.*, 2016: 8). Twitter now serves as a platform for news storytelling or 'produsage' (Bruns, 2008) through practices that combine the roles of the audience and the journalist, changing the dynamics of the public sphere. Twitter's feature of linking to other media situates it 'alongside the mass-mediated public' (Maireder and Asserhofer, 2014: 293). Traditional media organisations have embraced social media such as Twitter in their newsgathering and reporting practices, and news consumption across media is increasingly driven on content rather than media forms as users show 'declining loyalty' to certain types of media and outlets and instead forge personalised patterns of news consumption and sharing (Jensen *et al.*, 2016: 9).

In terms of engagement, danah boyd argues that the affordances of Twitter and other networked technologies facilitate the emergence of 'networked publics' that are simultaneously 'spaces constructed through networked technologies' and 'imagined collective[s] that emerge as a result of the intersection of people, technology and practice' (2010: 39). These publics function in the manner of offline publics in that they enable people to gather and connect for social, cultural and political purposes, but they are assembling and interacting in a unique way that is determined by networked technologies that reorganise information flows. The networked publics differ from traditional publics in the way they interact around issues, events and people and share information en masse.

Twitter enables users to share their multiple perspectives on political issues (Yardi and boyd, 2010) and apply hashtags as a way to frame and structure debates (Bruns, 2012). Organising content along themes and keywords, hashtags are 'a means of coordinating a distributed discussion between more or less large groups of users, who do not need to be connected through existing "follower" networks' (Bruns and Burgess, 2011b: 1). Thematic, topically focused hashtags are 'explicit attempts to address an imagined community of users' and have the potential to 'act as a bridge' between the hashtag community and the user's network (ibid.: 4). They create ad hoc channels, rather than groups. Unregulated and unclassified, hashtags are popular for their 'cultural generativity' (Burgess, 2011). In a media-saturated world, non-government organisations (NGOs) use news sharing as another strategy to increase visibility and agenda setting to engender social change in processes of 'trans-mediation'. In this process the raising of donations or mobilisation of protest 'link[s] the speech acts of protest reporting with the scenes of physical action' (Chouliaraki, 2013: 277).

Information-sharing on news and political topics, facilitated by functions such as 'Trending Topics', plays an agenda-setting function by marking out which issues are important at a given time. Twitter users create a news agenda that differs from the mainstream as content is 'filtered through the community's own established interests and news frames, resulting in a distribution of attention' (Bruns and Burgess, 2011a: 45). Therefore, Twitter can be understood as a networked public sphere in that its open and low-threshold manner of exchange encourages the sharing of public opinion and facilitates open public debate between potentially large numbers of users. Within these topical conversations, marked by hashtags and trending topics, influencers and gatekeepers can be identified, enabling the researcher to map the network topology and to explore hashtags as framing devices rather than merely 'lightweight, semantic annotations that publics assign to tweets in their efforts to self-tag generated content' (Meraz and Papacharissi, 2016: 104).

The Right to Food campaign

The Right to Food Coalition uses Twitter strategically, as a tool for outreach, communicating information and connecting individuals and organisations concerned with food security in Australia. It is employed to attain a level of publicness and publicity: to elevate the issue of domestic food insecurity within the public sphere. In Australia, a country recognised as one of the richest and most politically stable in the world, 1.2 million citizens in a population of 23 million are food insecure,[2] with Aboriginal and Torres Strait Islanders five times more likely than other Australians to be unable to feed themselves and their families on a regular basis. Individuals and families experiencing housing stress, those on low incomes, the elderly and the disabled are also likely to be unable to access or afford fresh healthy food (SecondBite, 2015).

The Coalition aims to raise public awareness of domestic food insecurity and the gaps in social provisioning. In doing so, it also aims to 'shift the debate from food poverty and charity to one demanding food as a fundamental human right' (Lindberg et al., 2016: 5). The right to food, a concept formally recognised by the Food and Agriculture Organisation of the United Nations (FAO, 2004), is not familiar or well recognised in First World economies such as Australia. The dimensions of the right to food include non-nutrient values such as cultural and dietary preferences and protection of the environment. Who will produce food, how, and for whose benefit? This rights-based framework places the onus on 'states and other actors [to] cooperate internationally to address structural impediments to fulfilling the right to food' (Ishii-Eiteman, 2009: 697).

The Coalition staged a focused Twitter campaign on the days 21–23 June in the lead-up to the 2016 Australian Federal Election. This campaign had two main aims: to elevate the issue of food insecurity among Australians by targeting elected members across a range of portfolios, both Ministers and their Shadows, and to increase the membership of the Right to Food Coalition. Members of the Right to Food Coalition with active Twitter accounts, including Red Cross Australia, Sustain (the Australian Food Network), the Youth Food Movement and the Asylum Seeker Resource Center, were contacted directly via email and encouraged to collectively pose a series of 'organisationally promoted frames'

(Hopke, 2015: 9) in the form of pre-fabricated questions, on specific days, to politicians via Twitter. These questions were:

- Why don't we regularly monitor or measure the number of people in Australia that struggle to put good food on the table? Join @right_to_food (Tuesday 21 June)
- Why do we talk about people not having affordable housing but not about people having affordable healthy food? Join @right_to_food (Wednesday 22 June)
- Why don't we talk about how much food insecurity costs society and how little we spend to address it? Join @right_to_food (Thursday 23 June)

Users were encouraged to use the hashtags #healthelection16, #healthelection, #auspol, and #ausvotes.[3] Suggested 'targets' included Standing and Shadow Minsters of portfolios including Health, Social Services, Families, Aging and Community. To attain maximum reach, users were urged to place a period before the targeted politician's Twitter handle to reach both sets of followers; this practice shifts a tweet from reply status to all the sender's followers. For example: '.@sussanley Why don't we regularly monitor/measure the no. of ppl in Aust that struggle to put good food on the table?Join @right_to_food'. It was also suggested that campaigners tweet images of themselves holding a piece of paper with the question and the handle @right_to_food.

Several actors who generated the hashtag #righttofood share not only an existing interest in food policy issues but also overlapping structural and communicative networks – that is, they are already 'following' one another on Twitter, and also working together on offline projects. For example, Red Cross and the Right to Food Coalition were co-sponsors, with St Vincent de Paul, St Mary's House of Welcome, Sacred Heart Mission and the Sydney Food Fairness Alliance, of a special edition of the Council to Homeless Persons' magazine *Parity* (2016), entitled 'Beyond emergency food: Responding to food insecurity and homelessness'. Red Cross Australia did not publicly participate in the campaign in line with its principles of neutrality and independence.

Frame analysis and the Twitter Capture and Analysis Tool Set (TCAT)

Framing, a concept developed to study the influence of mass communication messaging on audiences, is also useful in the study of online communication. Media discourse can be understood as a set of 'interpretative packages' (Gamson and Modigliani, 1989), each of which is structured by a frame or theme that connects the different semantic elements of a narrative (Pan and Kosicki, 1989). Frame analysis explores how patterns of discourse are established and how they mobilise publics. The verification of frames is made through the close study of campaign discourse, the 'stuff of framing' (Steinberg, 1998: 862). Erving Goffman (1974) first applied the concept of framing in studies of communicative acts, claiming that 'to frame is to select some aspects of a perceived reality and make them more salient in a communicating text, in such a way as to promote a particular problem, definition, causal interpretation, moral evaluation, and/or treatment recommendation' (Entman, 1993: 52).

Frames are deployed by journalists and activists to shape the public's understanding of how a problem comes to be, while introducing important criteria by which solutions should be evaluated. In his research on the dynamics of inter-organisational frame disputes within the nuclear disarmament movement in the 1990s, Robert Benford (1993) introduces a typology of dispute coding categories including diagnosis, prognosis and frame resonance. The first category refers to the need for activists to share a common interpretation of reality. This alignment of the individual frames, or 'schemata of interpretation' (Goffman, 1974), becomes the essence or *raison d'être* of a campaign as a process of issue construction. Prognosis requires agreement on what can be done to solve the problem that has been diagnosed, building an interpretive foundation that forms the basis of recruitment, participation and action. Finally, frame resonance is created through the strategic communication of the movement ideology. It may require compromise to attract large numbers of supporters or influential actors to the movement. Congruence with the experience, attitudes, culture and beliefs of the target public is essential if the frame is to mobilise action and therefore become motivational (Benford, 1993: 677; see also Entman, 1993).

Frame analysis has been applied to explore how food insecurity is socially constructed in high-income nations such as Canada (Mah *et al.*, 2014). Rein and Schön (1996) explore how actors define household food insecurity and identify its causes, consequences and solutions. Foremost is identifying responsible actors and the types of interventions required to fix the problem. There has been limited research on how food activists employ digital media to facilitate 'networked framing', described by Zizi Papacharissi as a 'process through which particular problem definitions, causal interpretations, moral evaluations, and/or treatment recommendations attain prominence through crowd-sourcing practices' (2015: 75).

Twitter user practices, including the ability to deploy hashtags and retweet, can facilitate frame setting and frame building (Meraz and Papacharissi, 2013). Citizen users on Twitter engage in networked framing with advocacy organisations such as the Right to Food Coalition as they revise, rearticulate and redisperse frames. Symbiotic, conversational interactions on topics of mutual interest between users on social media platforms facilitate the performance of identity (Goffman, 1959) and participation in processes of frame negotiation. This 'connective action' embodies self-motivated participation through the sharing of personal content that may be co-produced and co-distributed among peers, non-government organisations (NGOs) and other groups with congruent ideas, plans, resources and networks (Benkler, 2006; Bennett and Segerberg, 2012). The scalability of these networks, and their capacity to supplement traditional activist communication and linking practices, enables food security advocates like the members of the Right to Food Coalition to connect with other activists in diverse geographic locations and extend the reach of a campaign or movement.

In this study I applied the Twitter Capture and Analysis Tool Set (TCAT), a freely available platform of publicly available data comprised of millions of tweets, to build an issue inventory. TCAT provides 'robust and reproducible data capture and analysis', easy downloading and methodological transparency (Borra and Rieder, 2014: 263). I performed basic data processing to set up TCAT to extract URLs, hashtag #righttofood and @mentioned accounts @right_to_food belonging to the Right to Food Coalition, and @WePublicHealth, a curated Twitter account set up as part of an independent journalism project on public health issues. I then used TCAT to sort tweets posted over three days, from 21–23 June 2016, inclusive, into the following categories: straightforward tweets, replies/mentions and retweets. User statistics by individual, hashtag frequency, user mention frequency, URL, media and work frequency were also gathered.

Analysis of Twitter data can reveal networking topics, actors and media objects (Maireder and Asserhofer, 2014) and also lends itself well to textual analysis, including co-word analysis (Marres and Weltvrede, 2013). Identification of the URLs most shared in the period of analysis reveals networks of associations between media objects as mediators on issues, the relationships between discursive frames operating and 'subject drift' away from the original topic. A content-oriented perspective provides insight into shared meanings, ideologies, understandings, discourses and norms. Issue networks can be seen to merge via co-located hashtags – those occurring side by side in the same tweet – which provide subject matter classification (networking topics). Retweets identify significant posts and experts in the field (networking actors), while replies and followers facilitate network analysis. Unidirectional (one-way) and bidirectional (two-way) connections between individual users, media organisations and civil society groups can identify common interests, cliques, sub-conversations and spawning coalitions.

Results

The total number of tweets related to the Right to Food Coalition campaign over the three-day period was 249, of which 86 contained, collectively, 15 hashtags (Table 9.1). The most popular hashtag was #foodpoverty (52 mentions), followed by #ShareyourLunch (37), both generated by @foodpoverty, a civil society organisation in the United Kingdom with 1,773 followers, which is 'leading an independent campaign to ensure food poverty is not institutionalised' (foodpoverty, 2016). The next most frequent hashtag was #RightToFood (19) followed by #foodbanks, #hunger and #foodsecurity. The most prominent user name was @right_to_ food, the handle of the Right to Food Coalition (with 62 mentions), followed by @food poverty (20) and @FlavourCrusader (15), a highly active Australian user with 2,368 followers representing 'farmers, gardeners, cooks and eaters fighting for a healthy, fair and delicious food supply' (FlavourCrusader, 2016). As anticipated, given the instructions sent to the initiating users, Australian politicians also featured in the list of most frequently mentioned and were linked via the hashtags #healthcosts, #healthelection16 and #ausvotes.

Table 9.1 Most prominent hashtags and users, Right to Food Coalition campaign, 21–23 June 2016

Most prominent hashtags	Most prominent users
#foodpoverty (52 mentions)	@right_to_food (62)
#Shareyourlunch (37)	@foodpoverty (20)
#RighttoFood (19)	@FlavourCrusader (15)
#foodbanks	
#hunger	
#foodsecurity	
#healthcosts	
#healthelection16	
#ausvotes	

The networks of association between users, hashtags and media objects such as URLs illustrate a wide range of discursive frames at play, and demonstrate the synergies between local campaigns on a transnational scale. The Right to Food Coalition campaign mobilised Twitter users internationally. My geographic analysis of contributors to the Twitter conversation reveals a concentration of activity in Sydney and Melbourne but also strong interaction from UK cities including London and Liverpool. Many of these users took the opportunity to promote their own campaigns. The most commonly tweeted and retweeted URL was www.cancook.co.uk/share-your-lunch/, the website of the Merseyside-based Share Your Lunch organisation, Twitter handle @foodpoverty. The Right to Food Coalition's campaign also resonated with other UK-based social enterprises including East London Food Access (@ELFA), which promotes project FreshWell, which provides stalls of fresh produce in housing estates for 'those needing it most'.

> RT @ELFA_Ltd: FreshWell is going mobile! Want to support our next #foodpoverty venture? First thing: go to https://t.co/Zi5wuc5S36 and make a pledge.
> (FreshWell Mobile, 23 June 2016)

European engagement through FIAN International (the Food First Information and Action Network), based in Germany, was also identified. As the primary UN-recognised NGO campaigning on the right to food, FIAN's participation in the conversation plays an important role in establishing the significance of the right to food as an international concept. FIAN's handle @FIANista, along with @FlavourCrusader, received the majority of retweets. The single most retweeted post provided a link to FIAN's Facebook page, demonstrating synergy between the two social media platforms:

> RT @FIANista: Strengthening the #RightToFood movement at local level in #Nepal and empowering vulnerable communities https://t.co/pFsLxDlldO
> (FIAN, 21 June 2016)

For FIAN International, the framing of the right to food contests the corporate capture of food and nutrition governance at the expense of small-scale food producers, family farmers and eaters (Mann, 2014). This connects the Right to Food Coalition to the wider, international struggle for food sovereignty,[4] a related but distinct political project.

Discussion

The Right to Food Coalition made an effort to frame the conversation on food insecurity through prefabricated messages that were 'personalised, yet at the same time generic in nature' (Hopke, 2015: 9). Core activists generated the messages and hence the framings, engaging in a form of 'centralised strategic management' (Bennett and Toft, 2009: 252). These messages were amplified through organisational and individual supporter accounts including the @WePublicHealth curated Twitter account. Through @replies and retweets we can assess degrees of interaction and spaces for political conversation. For example, the data reveal 128 tweets with mentions, 89 retweets and a very limited number of replies (6), suggesting that users were more likely to retweet/share rather than engage in bidirectional interaction. Arguably, this demonstrates that most users accepted and were supportive of the political messages they received regarding the need for policy-makers to address the right to food of citizens in Australia and elsewhere.

While it is challenging to measure conversationality based on hashtags,[5] they provide a plurality of perspectives in the framing of issues. The hashtag #foodpoverty, for example, frames food insecurity as an economic problem. Commonly used in Britain to describe household food insecurity, food poverty is defined as 'insufficient economic access to an adequate quantity and quality of food to maintain a nutritionally satisfactory and socially acceptable diet' (O'Connor *et al.*, 2016: 432). The relationship between rising living costs and hunger has been the focus of studies in Canada where there is a tendency among development practitioners to believe that programmes aimed at reducing poverty will reduce food insecurity, yet 'while poverty is unquestionably a major contributor to food insecurity, not all the poor are hungry' (Agriculture and Agri-Food Canada, 2002: 59). With the increase in demand for food relief in Australia, and the subsequent increase in activity in the food charity and advocacy sectors, the term food poverty is becoming more popular in Australia to describe the 'wicked problem' (Levin, *et al.*, 2012) that is food insecurity.[6] While food security and food poverty share the principles of access, availability, utilisation and stability, 'food insecurity can exist without food poverty as a contributing influence however food poverty cannot exist without food insecurity' (O'Connor *et al.*, 2016: 431). Food insecurity is not simply a consequence of economic constraints, but can result from a range of factors including lack of cooking skills, equipment or basic knowledge about food preparation (Coe, 2013), as well as limited access to retail shops that offer healthy and affordable food options (Food Standards Agency, 2014). Through the #shareyourlunch campaign, the frame of food poverty has been 'sponsored' (Rein and Schön, 1996) by social enterprises, city councils in Liverpool, media outlets including @PositiveNewsUK and celebrity chefs like Simon Rimmer, and thus granted legitimacy which is problematic for advocates insisting on a more nuanced view of the problem that addresses the complexity of food insecurity, which is determined by factors including time, mobility, education and opportunities to eat socially (SecondBite, 2015).

The use of hashtags facilitates the cross-flow of information between organisations on multiple scales, contextualising the local but also widening the frame. Data and anecdotal evidence on the experience of food insecurity in different countries is circulated through the hashtag #righttofood, which discursively aligns local campaigns such as the Australian Right to Food Coalition, ELFA and FIAN. Twitter conversations bring these diverse issue publics together digitally in 'hybrid forums', defined as:

> open spaces where groups can come together to discuss technical options involving the collective, hybrid because the groups involved and the spokespersons claiming to represent them are heterogeneous, including experts, politicians, technicians, and laypersons who consider themselves involved … where questions and problems taken up are addressed at different levels in a variety of domains.
>
> (Callon *et al.*, 2001: 18)

The hybridity of these forums is fostered further when expert discourses combine with the more vernacular comments and insights of non-experts. For example, Fizz77 '#housing #London Views are my own, R/T not endorsement' tags Evie Copland, author of the blogpost #UKhousingfast.

> RT @ukhousingfast: Thanks for writing this personal and insightful blog about #foodpoverty @misscopland #ukhousingfast htt://t.co/8b62fwDxvH
>
> (Fizz77, 22 June 2016)

Instances of user-to-user communication are limited in this study and as such represent low levels of political conversation and participation. There is little evidence that the Right to Food Coalition's campaign has significantly engaged those experiencing household food insecurity in the conversation, and its impact on the targeted policy-makers is difficult to measure. What the campaign has revealed is a networked public of pre-existing advocates of the right to food. These actors represent 'emergent socio-political assemblages with shared or interlocking concerns who know themselves as, and act as, publics through media and communication' (Burgess and Matamoros-Fernández, 2016: 81). Thus, the local campaign has initiated a cross-flow of information between aligned movements while supporting activists' goals of developing a sense of collectivity, and connectivity, between organisations and individuals who share the goal of promoting the right to food in Australia.

Conclusion

This study has focused on tweets as sources of data and sites of the production of meaning. Bursty and ephemeral, Twitter is a 'noisy' yet conversational environment where 'successful exchanges can and do take place' (Honeycutt and Herring, 2009: 9; see also boyd, 2010: 10). Replies, mentions and retweets are conversational practice. They 'can knit together tweets and provide a valuable conversation infrastructure … whether participants are actively commenting or simply acknowledging that they're listening, they are placing themselves inside a conversation' (Honeycutt and Herring, 2009: 7). Further, Twitter architecture enables users to be aware of what others tweet about and permits them to access indirect information through the retweets they receive. Retweeting facilitates the diffusion of information, validates comments, expresses agreement and offers support for positions on important issues such as the right to food.

This study demonstrates how Twitter can be an important strategic resource for NGOs and other civil society groups. In the case of the Right to Food Coalition campaign, a limited number of organisations speak on behalf of those affected by food insecurity. The platform appears a resource for internal movement collective identity building and organising rather than a means to reach wider audiences. The community of contributors to #righttofood includes individuals, bloggers, researchers, social enterprises, NGOs and interest groups concerned about food insecurity around the world. The platform serves as a tool that enables advocates to connect, share perceptions, framings and strategies for tackling food insecurity in different contexts.

The plurality of meanings and orientations in the framing of the right to food through hashtags and other Twitter artefacts demonstrates their power to define an issue. Individually, personal tweets might seem 'insignificant', but viewed collectively, they might serve as 'a resource for future generations to understand life in the 21st century' (Raymond, 2010). In this case the dominance of the hashtag #foodpoverty reveals a challenge for activists who seek to reveal the complexity of food insecurity and attain a more specific grasp of the problem and its social and political determinants including access, distribution and affordability. Those most affected by food insecurity have much to contribute through sharing their personal stories in their own communities. Their voices are essential to capturing the subjective experience of those who are food insecure, and to develop a nuanced understanding of how hunger is socially constructed among the most developed economies.

This study is limited in its generalisability, and has not evaluated the success of the campaign in applying pressure on politicians to address the unacceptably high rates of food insecurity in Australia. Nor can a study of Twitter provide deep insight into public opinion regarding the issue, particularly when the platform is not frequented by, or is inaccessible to, those most food insecure. The medium specificity of social media interactions means they are not reliable as a single source of data on social issues (Burgess and

Matamoros-Fernández, 2016). An uncritical celebration of the achievements of digital media platforms limits the value of research as a 'critique of media power' (Atton 2008: 218). Accordingly, activist use of social media must be considered within the wider communication ecology. The critical media researcher must avoid identifying too closely with social media platforms as objects of study and also resist the assumption that all participants in campaigns share an undisputed collective identity. Rather, social media platforms can be considered sites for 'reflexive forms of activism' (Caroll and Hackett, 2006: 96), with the ability to involve advocacy organisations and ordinary citizens in new forms of political communication, as well as news production, distribution and consumption (Couldry, 2000). Future investigation of the interplay between different forms of mediation and a wide range of media practices and formats might reveal how advocates for the right to food can interact with media outlets, media technologies and media professionals in new strategic alliances (Mann, 2013).

In conclusion, better understanding of how Twitter conversations contribute to the development of traction or political influence on an issue can only be assessed through longitudinal studies and the application of multiple methods. Researchers are advancing the affordances of digital methods such as TCAT to study how issue publics emerge and consolidate through social media. In this way we can learn how digital media platforms and applications contribute to the shaping of public issues and the democratisation of public spheres. The challenge for digital methods is to take our analysis beyond the most dominant voices to those of previously marginalised stakeholders, and enable us to trace the complexity of public opinion emerging from generative publics on important societal issues such as food insecurity.

Notes

1 Web 2.0 is defined as a 'network as platform, spanning all connected devices' in an 'architecture of participation' that facilitates continuous improvement through consuming, remixing and sharing data from numerous sources, resulting in richer user experiences (O'Reilly, 2007: 17).
2 This figure was obtained by asking Australians whether they had run out of food, and were unable to afford to buy more, at least once in the previous 12 months (SecondBite, 2015).
3 This enduring hashtag originated in 2010, prior to an earlier election, as a central hashtag for political discussion (Bruns and Burgess, 2011a).
4 Food sovereignty is 'the right of each nation to maintain and develop its own capacity to produce its basic foods respecting cultural and productive diversity' (La Via Campesina, 1996). According to the people's movement, La Via Campesina, the right to food can only be realised when food sovereignty is guaranteed.
5 The asynchronous nature of Twitter, whereby sender and receiver are not necessarily present in the conversation at the same time, means exchanges may be separated by long periods of time (Small, 2011).
6 Wicked problems are characterised by a high degree of interdependencies, uncertainties, circularities and the conflicting agendas of multiple stakeholders that complicate efforts to develop solutions (Levin et al., 2012).

References

Agriculture and Agri-Food Canada (2002) *Canada's Second Progress Report on Food Security*. Ottowa: Agriculture and Agri-Food Canada.

Atton, C. (2008) Alternative media theory and journalism practice. In M. Boler (ed.) *Digital Media and Democracy: Tactics in Hard Times*. Cambridge, MA: MIT Press.

Benford, R. D. (1993) Frame disputes within the nuclear disarmament movement. *Social Forces*, 71(3): 677–701.

Benkler, Y. (2006), *The Wealth of Nations: How Social Production Transforms Markets and Freedom*. New Haven, CT: Yale University Press.

Bennett, W. L. (2004) Branded political communication: Lifestyle politics, logo campaigns, and the rise of global citizenship. On M. Micheletti, A. Follesdal and D. Stolle (eds) *Politics, Products and Markets: Exploring Political Consumerism, Past and Present*. New Brunswick, NJ: Transaction, pp. 101–125.

Bennett, W. L. and Segerberg, A. (2012) The logic of connective action. *Information, Communication & Society*, 15(5): 739–768.

Bennett W. L. and Toft A. (2009) Identity, technology, and narratives: transnational activism and social networks. In A. Chadwick and P. N. Howard (eds) *The Routledge Handbook of Internet Politics*. New York: Routledge.

Borra, E. and Rieder, B. (2014) Programmed method: Developing a toolset for capturing and analysing tweets. *ASLIB Journal of Information Management*, 66(3): 262–278.

boyd, d. (2010) Social network sites as networked publics: Affordances, dynamics and implications. In Z. Papacharissi (ed.) *Networked Self: Identity, Community, and Culture on Social Network Sites*. New York: Routledge, pp. 39–58.

Bruns, A. (2008) *Blogs, Wikipedia, Second Life, and Beyond. From Production to Produsage*. New York: Peter Lang.

Bruns, A. (2012) How long is a tweet? Mapping dynamic conversation networks on Twitter using Gawk and Gephi. *Information, Communication & Society*, 15(9): 1323–1351.

Bruns, A. and Burgess, J. E. (2011a) #ausvotes: How Twitter covered the 2010 Australian federal election. *Communication, Politics and Culture*, 44(2): 37–56.

Bruns, A. and Burgess, J. E. (2011b) The use of Twitter hashtags in the formation of ad hoc publics. In *Proceedings of the 6th European Consortium for Political Research (ECPR) General Conference 2011, University of Iceland, Reykjavik*. Brisbane: Queensland University of Technology, pp. 1–9.

Burgess, J. (2011) The iPhone Moment, the Apple Brand and the Creative Consumer: From hackability and usability to cultural generativity. In L. Hjorth, J. Burgess and I. Richardson (eds) *Studying Mobile Media*. London: Routledge.

Burgess, J. and Matamoros-Fernández, A. (2016) Mapping sociocultural controversies across digital media platforms: one week of #gamergate on Twitter, YouTube, and Tumblr. *Communication Research and Practice*, 2(1): 79–96.

Callon, M., Lascoumes, P. and Barthe, Y. (2001) *Acting in an Uncertain World: An Essay on Technical Democracy*. Cambridge, MA: MIT Press.

Caroll, W. and Hackett, R. (2006) Democratic media theory through the lens of social movement theory. *Media, Culture and Society*, 28(1): 83–104.

Carr, D. (2012) Hashtag activism and its limits. *New York Times*. Available at: www.nytimes.com/2012/03/26/business/media/hashtag-activism-and-its-limits.html (accessed 22 January 2016).

Castells, M. (2009) *Communication Power*. New York: Oxford University Press.

Castells, M. (2012) *Networks of Outrage and Hope: Social Movements in the Internet Age*. Cambridge: Polity.

Chouliaraki, L. (2013) *The Ironic Spectator: Solidarity in the Age of Post-Humanitarianism*. Cambridge: Polity.

Coe, S. (2013) Feeding the family, are food prices having an effect? *Nutrition Bulletin*, 38(3): 332–336.

Cottle, S. and Lester, L. (2011) *Transnational Protest and the Media*. New York: Peter Lang.

Couldry, N. (2000) *The Place of Media Power: Pilgrims and Witnesses of the Media Age*. London: Routledge.

Critz, M. (2010) It's critical the future generations know what flavour burrito I had for lunch. Web log post. 28 April. Available at: http://blogs.loc.gov/loc/2010/04/the-library-and-twitter-an-faq/ (accessed 16 June 2016).

Entman, R. M. (1993) Framing: Towards clarification of a fractured paradigm. *Journal of Communication*, 43(4): 51–58.

Fizz77 (2016) RT @UKhousingfast [Twitter], 22 June. Available at: https://eviecopland.wordpress.com/2016/06/22/ukhousingfast/ (accessed: 21 December 2016).

FlavourCrusader (2016) We're farmers, gardeners, cooks and eaters fighting for a healthy, fair and delicious food supply. Join us. [web log post]. Available at: https://twitter.com/FlavourCrusader? lang=en (accessed 2 December 2016).

Food and Agriculture Organisation of the United Nations (FAO) (2004) Voluntary guidelines to support the progressive realisation of the right to adequate food in the context of national food security. Available at: www.fao.org/docrep/009/y7937e/y7937e00.htm (accessed 15 September 2017).

Foodpoverty (2016) Leading an independent campaign to ensure food poverty is not institutionalised. We are about solutions, dignity, food quality and choice. [on Twitter], 23 June. Available at: https://twitter.com/search?q=%40foodpoverty&src=typd&lang=en (accessed 23 June 2016).

Food Standards Agency (2014) Food poverty. Available at: www.food.gov.uk/northern-ireland/nutritionni/ninutritionhomeless (accessed 24 December 2016).

Freshwell Mobile (2016) Spacehive, [on Twitter], 23 June. Available at: www.spacehive.com/freshwell-mobile (accessed 23 June 2016).

Fuchs, C. (2014) *Social Media*. London: SAGE Publications.

Gamson, W. A. and Modigliani, A. (1989) Media discourse and public opinion on nuclear power: A constructionist approach. *American Journal of Sociology*, 95: 1–37.

Goffman, E. (1959) *The Presentation of Self in Everyday Life*. Garden City, NY: Doubleday.

Goffman, E. (1974) *Frame Analysis: An Essay on the Organisation of Experience*. Cambridge, MA: Harvard University Press.

Guidry, J. A. and Sawyer, M. Q. (2003) Contentious pluralism: The public sphere and democracy. *Perspectives on Politics*, 1: 273–289.

Habermas, J. (1989) *The Structural Transformation of the Public Sphere*. Cambridge, Polity Press.

Habermas, J. (2006) Political communication in media society: Does democracy still enjoy an epistemic dimension? The impact of normative theory on empirical research. *Communication Theory*, 16: 411–426.

Honeycutt, C. and Herring, S. C. (2009) Beyond microblogging: conversation and collaboration via Twitter, in *Proceedings of the Forty-Second Hawai'i International Conference on System Sciences*. Big Island, HI: IEEE Press, pp. 1–10.

Hopke, J. E. (2015) Hashtagging politics: Transnational anti-fracking movement Twitter practices. *Social Media and Society*, July–December: 1–12.

Ishii-Eiteman, M. (2009) Food sovereignty and the international assessment of agricultural knowledge, science and technology for development. *Journal of Peasant Studies*, 36(3): 689–700.

Jensen, J. L., Mortensen, M. and Ørmen, J. (eds) (2016) *News across Media: Production, Distribution and Consumption*. New York: Routledge.

Keck, M. and Sikkink, K. (1998) *Activists Beyond Borders: Advocacy Networks in International Politics*. Ithaca, NY: Cornell University Press.

Kwak, H., Chun, H. and Moon, S. (2011) Fragile online relationship: A first look at unfollow dynamics in Twitter. In *Proceedings of CHI 2011*. 7–12 May. Vancouver, BC, Canada. Available at: http://an.kaist.ac.kr/~haewoon/papers/2011-chi-unfollow.pdf (accessed 5 January 2016).

La Via Campesina (1996) The right to produce and access to land, declaration 11–17 November 1996. Available at: nyeleni.org (accessed 1 May 2017).

Levin, K., Cashore, B., Bernstein, S. and Auld, G. (2012) Overcoming the tragedy of super wicked problems: Constraining our future selves to ameliorate global climate change. *Policy Sciences*, 45(2): 123–152.

Lindberg, R., Kleven, S., Barbour, L., Booth, S. and Gallegos, D. (2016) Introduction. In *Parity: Beyond Emergency Food: Responding to Food Insecurity and Homelessness*, 29(2): 4.

Mah, C., Hamill, C., Rondeau, K. and McIntyre, L. (2014) A frame-critical policy analysis of Canada's response to the World Food Summit 1988–2—8. *Archives of Public Health*, 72(41): 1–7.

Maireder, A. and Asserhofer, J. (2014) Political discourses on Twitter: networking topics, objects and people. In K. Weller, A. Bruns and J. Burgess (eds) *Twitter and Society*. New York: Peter Lang, pp. 305–318.

Mann, A. (2013) Bursting the 'Brussels Bubble': The movement towards transparency on European farm subsidies. *Ethical Space: The International Journal of Communication Ethics*, 10(2/3): 47–54.

Mann, A. (2104) *Global Activism in Food Politics: Power Shift*. Basingstoke: Palgrave Macmillan.

Mann, A. (2016) The right to food and how 1.2 million Australians miss out. *Croakey*, web log post, 10 April 2016. Available at: https://croakey.org/the-right-to-food-and-how-1-2-million-australians-miss-out/ (accessed 15 November 2016).

Marres, N. and Rogers, R. (2005) Recipe for tracing the fate of issues and their publics on the web. In B. Latour and P. Wiebel (eds) *Making Things Public: Atmospheres of Democracy*. Cambridge, MA: MIT Press, pp. 922–935.

Marres, N. and Weltevrede, E. (2013) Scraping the social: Issues in real-time social research. *Journal of Cultural Economy*, 6(3): 313–335.

Meraz, S. and Papacharissi, Z. (2013) Networked gatekeeping and networked framing on #Egypt. *The International Journal of Press Politics*, 18(2): 138–166.

Meraz, S. and Papacharissi, Z. (2016) Networked framing and gatekeeping. In T. Witschge, C. W. Anderson, D. Domingo and A. Herminda (eds) *The SAGE Handbook of Digital Journalism*. London: SAGE, pp. 95–112.

Miller, V. (2008) New media, networking and phatic culture. *Convergence*, 14(4): 378–400.

Morozov, E. (2009) Foreign policy: Brave new world of Slacktivism. *NPR*, 19 May. Available at: www.npr.org/templates/story/story.php?storyId=104302141 (accessed 15 October 17).

O'Connor, N., Farag, K. and Baines, R. (2016) What is food poverty? A conceptual framework. *British Food Journal*, 118(2): 429–449.

Olesen, T. (2005) Transnational publics: New spaces of social movement activism and the problem of global long-sightedness. *Current Sociology*, 53(3): 419–440.

O'Reilly, T. (2007) What is Web 2.0?: Design patterns and business models for the next generation of software. *Communications & Strategies*, 65: 17–37.

Pan, Z. and Kosicki, G. M. (1989) Framing analysis: An approach to news discourse. *Political Communication*, 10(1): 55–75.

Papacharissi, Z. (2015) *Affective Publics: Sentiment, Technology, and Politics*. New York: Oxford University Press.

Parity: Beyond Emergency Food: Responding to Food Insecurity and Homelessness (2016), vol. 29, no. 2.

Raymond, M. (2010) The library and Twitter: An FAQ. *Library of Congress Blog*, web log post, 28 April 2010. Available at: http://blogs.loc.gov/loc/2010/04/the-library-and-twitter-an-faq/ (accessed 20 October 2016).

Rein, M. and Schön, D. A. (1996) Frame-critical policy analysis and frame-reflective policy practice. *Knowledge Policy*, 9(8): 5–104.

Right to Food Coalition (2016) The right to eat well no matter where you live. Available at: https://righttofoodcoalition.files.wordpress.com/2016/04/rtf-final_food-access-in-aus_18-april-2016.pdf (accessed 20 October 2016).

Rogers, R. (2009) *The End of the Virtual: Digital Methods*. Amsterdam; Amsterdam University Press.

Rosen, J. (2006) The people formerly known as the audience. *Press Think: Ghost of Democracy in the Media Machine*. Available at: http://archive.pressthink.org/2006/06/27/ppl_frmr.html (accessed 20 October 2016).

Rousseau, S. (2012) *Food and Social Media: You Are What You Tweet*, Plymouth: AltaMira Press.

Ruiz, P. (2014) *Articulating Dissent: Protest and the Public Sphere*. London: Pluto Press.

Sarno, D. (2009) Twitter creator Jack Dorsey illuminates the site's founding document. Part 1. *Los Angeles Times*, 18 February 2009, Available at: http://latimesblogs.latimes.com/technology/2009/02/jack-dorsey-on.html (accessed 15 October 2016).

SecondBite (2015) *Food Security: What Is It? Fact Sheet*. Available at: http://secondbite.org/sites/default/files/SecondBite%20Fact%20Sheet%20Series_Food%20Insecurity.pdf (accessed 12 January 2017).

Small, T. A. (2011) 'WHAT THE HASHTAG?' A content analysis of Canadian politics on Twitter. *Information, Communication & Society*, 14(6): 872–896.

Socialbakers (2016) *Twitter Statistics Directory*. Available at: www.socialbakers.com/statistics/twitter/ (accessed 13 November 2016).

Steinberg, M. (1998) Tilting the frame: Considerations on collective action framing from a discursive turn. *Theory and Society*, 27(6): 845–872.

Turoff, M. and Hiltz, S. R. (2009) The future of professional communities of practice. In C. Weinhardt, S. Luckner and J. Stosser (eds) *Designing e-business Systems: Markets, Services, and Networks*. Berlin: Springer, pp. 144–158.

Weller, K., Bruns, A. and Burgess, J. (2014) *Twitter and Society*. New York: Peter Lang.

Yardi, S. and boyd, d. (2010) Dynamic debates: An analysis of group polarization over time on Twitter. *Bulletin of Science, Technology & Society*, 30(5): 316–327.

10 Food politics in a digital era

Tania Lewis

Introduction

A farmer on a small-scale organic farm in rural India uploads images of his latest produce to consumers and retailers via an open source online food hub; a 'conscious consumer' in Charlottesville, Virginia, uses an app while supermarket shopping to look up the ethical credentials of food producers and product ingredients; a business woman on a work trip to Rio de Janeiro is 'informed' by a travel and food app on her smartphone where and what she might like to eat for breakfast based on GPS technology and her previous preferences.

These three diverse examples speak to the changing nature of our engagements with food today in an increasingly digital world. From home cookery to restaurant going, from farming to food politics, the world of food is being quietly colonised by an array of electronic devices, online content and information and communication technologies. Meanwhile the realm of the digital has become increasingly concerned with all things food-related, from endless food snapshots on Facebook and Instagram to the rise of YouTube cooking and food channels, the fastest-growing genre on the video-sharing service.

This digital 'turn' in the lives of many people has unsurprisingly inspired a huge amount of commentary and reflection, some of it celebrating the capacity of technology to connect and empower us, while other accounts offer a more dystopian vision of data control and surveillance (Andrejevic, 2007). Numerous fine-grained studies have sought to engage with the emergent role of the digital in shaping our everyday domestic lives, interpersonal relations and consumer practices while a number of larger-scale accounts have emerged examining the digitisation of work, society and politics, governance, democracy and citizen engagement (Terranova, 2004; Papacharissi, 2010; Van Dijck, 2013). Yet, surprisingly little has been written on the growing entanglements between the digital and the world of food, and particularly the rise of new forms of political engagement and food citizenship (exceptions include Eli *et al.*, 2016; Humphery and Jordan, 2016; and Lupton, forthcoming).

Over the past couple of decades, we have seen the media – from popular documentaries to reality cooking shows to foodie lifestyle shows – increasingly concerned with political and ethical questions of how we consume in the Global

North (Lewis, 2008, 2012; Lewis and Potter, 2011). This has been accompanied by the rise of some fairly unlikely spokespeople for issues of food politics including celebrity chefs and even celebrity farmers. Questions about how our food is produced and how sustainable imported food might be, as well as debates about animal welfare and genetically modified organisms (GMO), have become the fodder of both television and film.

Increasingly, however, these questions and concerns have extended into the digital realm. In the context of a growing emphasis on personal and household practices of consumption and lifestyle as potential sites of political and ethical engagement with food (Goodman *et al.*, 2010; Lewis and Potter, 2011), the particular affordances offered up by digital technologies have begun to shape a whole new terrain of engagement and activism around food; not only for consumers and the odd digitally-empowered celebrity chef, but for an array of other actors, from farmers and alternative food movements to powerful corporate food players (Peekhaus, 2010; Choi and Graham, 2014; Bos and Owen, 2016).

The realm of digital food politics is a vast emergent field. In this chapter, I will touch upon just some of the themes and concerns facing food citizens, producers and activists. In offering a broad overview of some of the key issues raised by the digitisation of food politics, the chapter is structured accordingly: the first section briefly outlines the growth of lifestyle and consumer-related forms of participatory politics online, while the next section looks at the affordances of online platforms for enabling connected forms of personal consumption. The third section discusses the role of platforms in bringing together food communities. This is followed by a discussion of the limitations of connectivity and the hegemony of corporate food politics in social media spaces. The conclusion discusses the limits of data sharing around food and so-called informational transparency in an era of data monitoring and 'Big Data'.

Consumption and food politics: participatory personal politics online

In their book *YouTube: Online Video and Participatory Culture*, Jean Burgess and Joshua Green suggest that digital media engagement is no longer about consumers and producers, professionals and amateurs, non-commercial versus commercial players, but instead needs to be understood in terms of 'a continuum of cultural participation' (2009: 57). Enabled by the affordances of Web 2.0 and by video conversion and sharing technology, websites such as YouTube have heralded a dramatic shift from the digital consumer as passive downloader to an upload-and-exchange culture of creativity and pro-sumerism.

In the space of food politics, the rise of the consumer as a new powerful actor enabled by digital resources speaks in interesting ways to what Juliet Schor (1999) has termed 'the new politics of consumption' whereby everyday acts of purchasing and consuming food, goods, energy and water become tied to a sense of global responsibility. For the reflexive late-modern self, once-ordinary everyday practices become imbued with both biographic and broader social and political significance or what Giddens (1991) has referred to as 'life politics'. The question then becomes: what forms of political civic engagement are we now seeing in a digital realm in which the users are seemingly enabled to make their voices heard as individuals but also to connect at a larger level with virtual communities?

There has been considerable critique of the rise of the 'intimate' forms of 'political engagement' we might associate with the digital realm. American cultural theorist Lauren Berlant, in her work on intimate publics, talks rather scathingly of the way in which the centre of political life has shifted towards the private sphere with citizenship increasingly seen as being 'produced by personal acts and values' (1997: 5). This shift, for Berlant, constitutes '[d]ownsizing citizenship to a mode of voluntarism' (ibid.: 5). In contrast, Swedish political scientist Michele Micheletti argues that political consumerism does not crowd out other, more traditional forms of civic engagement, but rather promotes what she sees as 'individualized collective action' (2003).

Whether we see these developments in positive or negative terms, it is clear that the digital realm is a space in which a growing range of forms of political participation is taking place. As José van Dijck argues in *The Culture of Connectivity*: 'Within less than a decade, a new infrastructure for online sociality and creativity has emerged, penetrating every fiber of culture today' (2013: 4) – and, in turn, shaping the landscape of civic and political engagement in a range of unforeseen and unpredictable ways. Here I discuss some of these new and emergent practices, from consumer-driven online activism to digitalised alternative food networks, in ways that move beyond celebratory or dystopian images of online participation. This chapter aims to demonstrate that the digital realm offers a wide range of potential political affordances that cannot necessarily be pre-empted in advance as progressive, conservative or otherwise. My own research on the Permaculture movement (discussed in more detail below), for instance, indicates that the Internet is providing food movements with platforms for connecting individuals and households in ways that may significantly boost the impact and visibility of these movements (Lewis, 2015). On the other hand, the connective capacity of social media, while enabling new forms of food activism, can also be seen as extending the marketing and surveillance power of major corporations and the state to hitherto unimagined degrees (Suarez-Villa, 2016). Nick Couldry thus argues, in his call to understand and interrogate social media platforms *as institutions*, that the utopianism of connected sociality needs to be recognised as 'a playground for deep economic battles about new forms of value, value generated from data, the data that we generate as we act online' (2015: 641).

Digital tools and platforms: connected consumption and apptivism

While digital social relations are intrinsically shaped and constrained by what we might term 'compromised connectivities', Couldry's (2015a) metaphor of the playground speaks to the ways in which digital socialities are underpinned by complex interplays of contestation and hegemony. As Deborah Lupton suggests in her recent book *Digital Sociology*, '[d]igital technologies have created new political relationships and power relations' (2015: 189). This is evident in the increasing power struggles over information, data, visibility and transparency that mark the contemporary digital foodscape and the rise of digital forms of 'food citizenship' (Booth and Coveney, 2015; Gómez-Benito and Lozano, 2014; Wilkins, 2005).

Here I discuss how the affordances of the Internet offer potential for new forms of not only consumer literacy about the politics of food consumption but also connectivity. A key shift associated with the move to digital platforms for consumers is the connection that everyday digital devices and platforms make with people's actual practices, whether when they are making food choices at the grocery store or market, or when they are sharing their food skills and knowledge with others.

Numerous platforms are concerned with virtually connecting people's everyday food practices to processes of global food production. *National Geographic*'s interactive food documentary 'A five-step plan to feed the world' (Foley, 2017) offers a re-visioning of the usual depiction of the elongated exploitative commodity chain associated with agri-business, reconnecting us with the 'faces' of small farmers around the world whom they see as the new drivers of a green agricultural revolution. Ethical shopping websites like *Follow the Things*,[1] on the other hand, make visible and materialise the politics of production behind the 'commodity chains' or 'networks' that bring us Israeli avocados grown on illegally seized Palestinian land or bananas grown by Nicaraguan banana workers who have sued Dole, one of the biggest food corporations in the world, for exposing them to a banned pesticide linked to severe health problems – though some of this information is dated, illustrating the contrast between a static website with a relative lack of updates versus the more dynamic flows of information associated with crowdsourced apps.

A key area here is the growing use of apps, or what has been termed 'apptivism', to enable more reflexive critical approaches to consumption. In the context of the contemporary globalised foodscape, there has been a growing public concern about the disconnection between consumers and the origins of their food – from questions about health and food safety, to concerns about animal welfare and the exploitation of Third World farmers and producers (Crang and Cook, 2004; Coff *et al.*, 2006; Goodman *et al.*, 2010; Lewis and Huber, 2015). In response to these concerns, a large range of food apps has emerged, attempting to create conditions of what we might term 'connected consumption'. Such apps seek to reconnect consumers to their food through various means, whether enabling individual consumers to access information about the origins of a food product's ingredients via a barcode scanning app or through linking consumers to a broader community of ethical consumer citizens.

The Good Guide, one of the most developed online ethical shopping guides, represents the former, information-focus kind of app. Employing a team of researchers including environmental scientists and sociologists, it offers a range of technological quick fixes for ethical consumers on the go, including a mobile phone app that allows you to use barcode technology to get just-in-time information about the environmental, health and social impacts of companies and products. The Good Guide team describes this app as like having 'a PhD in your pocket' (Viswanathan, 2010). They also offer a 'personal filter' option on commodities whereby you can tailor the guide's app to fit your own personal ethical concerns and lifestyle needs (i.e. you might tick the box for animal cruelty-free products but not for products that minimise their impact on the environment and/or support labour rights), arguably reducing complex interconnected issues around the global ethics and politics of consumption to purely a question of personal taste.

Rather more interesting are those apps that attempt to shift people's thinking away from shopping per se and towards other forms of exchange. One example is OLIO, an app that aims to connect neighbours with each other and/or with local businesses to exchange any surplus food they might have, with app users able to share images of unused food with the OLIO community. On its website, OLIO also emphasises the potential benefits of the app for businesses, noting that the act of sharing leftover food can bring in new customers, reduce food waste and help business 'connect with your community' (OLIO, 2016).

OLIO emphasises a civic-minded digital consumerism with a commercial edge where the act of sharing leftover food is framed as a potentially collective and transformative act ('a food sharing revolution') but with potential personal and business benefits. If value-driven acts of consumption have become a new stage for enactments of civil society and political agency, then the just-in-time, connected affordances of personalised digital apps like OLIO are in the vanguard of the new politics of consumption.

However, there are obvious limits to this kind of privatised, consumer-driven approach to 'revolutionising' food practices. Many of these apps and websites offer a fairly banal ethical 'shopping can save the world' message without necessarily challenging the foundations of commoditised food consumption itself. Environmentalist and *Guardian* journalist George Monbiot comments rather scathingly about such forms of caring consumption: 'It does not matter whether we burn fossil fuels with malice or with love' (2007: 112). Further, as Berlant's critique of intimate politics suggests, there is a problematic tendency to privatise, and therefore to potentially minimise, issues that are of major public or global concern. Such practices, while well meant, may thus deflect from rather than contribute to changing the reality of global agri-business and the inequities of today's urban foodscapes by offering a quick and easy ethical fix for privileged consumers and business owners who have the choice to share and distribute food.

However, the digital space has also seen the emergence of more critical uses of interactive platform technology that aim to connect a range of actors and groups concerned with offering genuine alternatives to global agri-business. In the next section, I discuss the ways in which online platforms have enabled a range of forms of alternative political organisation and community building in the critical food domain.

Crafting and connecting food communities online

> Although it may seem the most unlikely of catalysts, digital technology is jogging our memories of *real food* and *agrarian culture* [my emphasis]. We may be going back to the land, but many of us are bringing our laptops and smart phones.
>
> (Hatfield *et al.*, 2008: 4)

As the quote from Hatfield and colleagues suggests, somewhat paradoxically, the digital realm is playing a key and growing role in linking producers to 'authentic' food and to 'the land,' enabling a sense of *reconnection* with where food comes from. For small-scale food producers in particular, Web 2.0 allows them to connect with consumers directly and also with each other, thus potentially facilitating alternatives to globalised food and the elongated commodity chains of agri-business (Hearn *et al.*, 2014).

Another key area in which digital connectivity enables significant alternative forms of political organisation and community building is in the form of food networks. For instance, while still a nascent area of research, critical scholarship has begun to emerge on the roles of social media in shaping and enabling various virtual, networked alternatives to globalised food (ibid.). In their article, 'The online spaces of alternative food networks in England', Elizabeth Bos and Luke Owen explore the way in which the digital realm offers opportunities for reconnection with the 'complex systems of food provisioning' that have worked 'to distance and disconnect consumers from the people and places involved in contemporary food production' (2016: 1). Studying eight alternative food networks and 21 online spaces, they examine the ways that producers and citizens alike are using digital means to reconnect with local and rural food production to build potential alternatives to global agri-business. As their research on online food hubs suggests, digital networks are as much about strengthening connections to place and the local, as they are about forging global links.

One example of this is the Open Food network (openfoodnetwork.org) – an original Australian online collective, which is based on open source technology developed by US food hubs – Local Dirt and the Oklahoma Food Coop[2] (Open Food Network 2015; Local Dirt, 2016; Oklahoma Food Cooperative, 2016). Linking consumers, food hubs and farmers to local food enterprises, the Open Food network has been taken up as a model around the world.

Web 2.0 has also enabled forms of what Bennett and Segerberg (2013) have called 'connective action' for a range of other kinds of 'communities' including private householders seeking to collectivise their personal 'experiments' in living.

As noted above, another generative example of a highly localised group of urban food production activists who have employed digital technology is Permablitz Melbourne,[3] a group with whom I've conducted participatory ethnographic research and whom I want to discuss in some detail here. A group of volunteers who conduct makeovers in ordinary suburban gardens throughout Melbourne following the principles of permaculture,[4] Permablitz Melbourne is not so much a community in the conventional sense as a fluid network of people, connected primarily via a website. The Permablitz Melbourne case foregrounds in interesting ways how even the most strongly locally rooted forms of food activism have become co-articulated with the digital realm. Since the Melbourne network formed, Permablitz groups have emerged in Sydney, Adelaide, Alice Springs, Darwin, Canberra, Tasmania, the Sunshine Coast, California, Montreal, Istanbul, Jogjakarta, Bali, Uganda and beyond. While Permablitz is now internationally connected via the global mobility of key knowledge brokers and trained Permaculturalists, with the movement made visible and linked internationally through a strong web presence, membership of the Permablitz community is also a very hands-on and localised affair, as an excerpt from my fieldnotes from a Permablitz I participated in a couple of years ago suggests.

> I arrived somewhat late in the morning to the Sunday 'blitz', travelling to a part of northern Melbourne I hadn't visited before. Up above the Pentridge jail housing redevelopment into a tiny street just off the mixed industrial northern reaches of multicultural Sydney Rd. The telltale signs of a small suburban cul de sac chock-a-block with cars suggested something was going on behind the blankfaced houses. Armed with a shovel, hat and sunscreen I followed a lanky stranger down the side drive of an ordinary brick house to find a good sized group of people already at work weeding, hacking away at plants and thoughtfully inspecting the various spaces and 'projects' underway in the to my (inner urban) eyes rather huge quarter acre block. Cut to the end of the day and I and others, no longer strangers, are taking photos and videos of the transformation to post on the Permablitz website. As our collective images will show, in one day with the aid of planning and the labour and skills of many bodies, a large neglected suburban backyard is on its way to turning into an integrated permaculture garden complete with chickens, an urban farm in the making.

My fieldnotes foreground the at once privatised yet communitarian dimensions of this movement. 'Blitzes' are disparate events (usually occurring over one day, though with a longer lead time in terms of planning and preparation, see Figure 10.1) that take place in people's backyards around suburban Melbourne. The participants, most of whom are strangers, are put in touch with the particular Permablitz organiser in question through a website, often travelling long distances across town to help with the permablitz. While the core members of permablitz are foundational to its identity as a movement and meet regularly, for the most part, digital technology is key to the formation and maintenance of this 'community'.

In the case of Permablitz Melbourne, as a collective of essentially like-minded strangers distributed across the city, many of whom may never actually meet, digital technology is pivotal to 'connecting' members of the

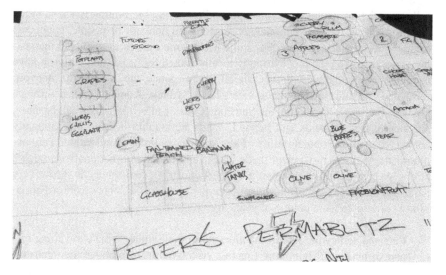

Figure 10.1 Plans for a permaculture garden.

organisation and more fundamentally to organising the citywide backyard blitzes that are integral to the movement. With no bricks and mortar 'club-house', Permablitz Melbourne's 'home' is essentially a website, www.perm-ablitz.net (accessed 15 October 2017), with members' back gardens representing ad hoc meeting places for conducting blitzes and holding work-shops. During and around the time of the permablitz, the garden in question thus shifts from being a private place to a public and political space of sorts, a resource for the permaculture and broader 'community' in terms of both being a site of potential food production (reflecting the ethos of the website's logo 'eating the suburbs – one backyard at a time'), but also a space in which skills and expertise are shared with and passed on to other 'blitzers' via the work-shops and talks that are held on the day.

Photos, and in some cases time-release videos, of the transformational process are also posted to the website after each permablitz, serving as a way of docu-menting and publicising these various backyard endeavours as well as an adver-tisement to attract potential future blitzers, while also linking Permablitz Melbourne to other Australian and international permablitz and permaculture communities and offering a guide on how to set up a regional permabitz network.

While driven by households and highly localised experts and intermediaries in the first instance, the Permablitz community and the processes of backyard and suburban transformation that the community and its members enable are joined up and 'public-ised' via digital platforms and digital networks, making it a good example of how virtual connectivity can help dispersed groups of actors to engage in a kind of Michelettian 'individualized collectivism' (Micheletti, 2003).

Another key generative aspect of the connective action enacted through systems like the Open Food Network and Permablitz is the way it raises questions of *scale*. The development of localised and regional food systems has often been hampered by problems of scale, particularly how to collectivise efforts and scale up while still adhering to social and environmental drivers. Here the connective affordances of web platforms have greatly bolstered the ability of distributed local systems to expand and collectivise. As I have argued elsewhere in relation to Permablitz Melbourne and the potentially transformative impact of other 'local' sustainability enterprises:

> The cascading effects of seemingly small-scale initiatives, such as the role played by local and personal actions in the Arab spring, often in conjunction with networked media, have foregrounded the increasing insufficiency of macro and micro conceptions of scale in relation to theorizing political activism, engagement and transformation.
>
> (Lewis, 2015: 351)

Complicating connectivity: corporate food politics online

While Web 2.0 has undoubtedly greatly enabled alternative food networks to globally connect and scale up, for many everyday web users, the main engagement with the digital realm is through what Zizi Papacharissi characterises as 'commercially public spaces' (2009: 242). The digital realm is not only largely owned by commercial interests but its infrastructure and data logics are also thoroughly monetised. In the food domain, the world's top agricultural and fast food companies are all significant adopters of social media (Stevens *et al.*, 2016) while Twitter, Facebook and Instagram draw a significant proportion of their revenue from corporate-funded online advertising.

Controversial agricultural mega-corporations such as Monsanto, described by Bloomberg as America's 'Third-most-hated company' (Bennett, 2014), have invested heavily in a social media presence. For instance, Monsanto has sought to counter criticism from anti-GMO activists by mounting a 'be part of the conversation' campaign on social media in which members of the public are invited to ask questions about the company and its practices. Rebranding itself via Twitter using discourses of 'sustainable development' and 'biological conservation', Monsanto[5] has attempted to position itself as a responsible corporate player engaged in a 'rational' public dialogue around nutrition, sustainable population growth and environmentalism (see Peekhaus, 2010; Monsanto, 2015).

Monsanto's use of Twitter to position itself in an 'open' dialogue with the community highlights the way in which social media are associated with a seductive 'myth of natural collectivity' (Couldry 2015b: 620). While the tweets of teenagers and those of global corporate CEOs apparently receive equal airtime in the democratic realm of social media, as Couldry contends, the new media landscape is nevertheless dominated by corporations that strive to control and above all monetise the everyday connective practices of web users.

While the rise of alternative food networks online highlights the power of connective action, the flipside of virtual engagement then is the way in which connectivity itself has become a key source of value for corporate players through both audience labour and the traces of data produced by acts of connectivity. Whether through the act of demonstrating one's 'alternative' lifestyle and consumer preferences through 'likes' on Facebook or by producing videos for YouTube demonstrating backyard permaculture, digital audiences can be seen as effectively providing 'free' labour which can be potentially exploited by capital. As van Dijck explicates, while social media emerged out of a participatory ethos of creativity and exchange: 'Connectivity quickly evolved into a resource as engineers found ways to code information into algorithms that helped brand a particular form of online sociality and make it profitable in online markets' (2013: 4).

People's everyday engagements and practices with digital media can be commoditised and exchanged between digital media companies and advertisers (Fuchs, 2013). This subtle and often hidden commercial exploitation of digital interactivity can be seen as part of a broader shift to the 'experience economy' (Pine and Gilmore, 1999; World Economic Forum, 2016), a shift that marks a transition from consumers buying goods to a growing focus on *doing*. Increasingly corporates are attempting to tap into this turn towards the experiential. One example in the household food realm is the rise of 'free' services such as Whatscook which is sponsored by the mayonnaise manufacturers Hellmann's. Purportedly assisting hapless householders to cook up convenient meals based on what's in their fridge (through using WhatsApp to connect people to Hellmann chefs), the app also enables the collection of household data on food and lifestyle preferences marking what I would argue is a broader monetisation and corporate surveillance of the so-called 'social' exchanges in social media.

Thus, aside from the labour that users provide as they interact with and upload data online, another key area where the digital realm has converted audience engagement into a resource is through the generation of *data*. As Lupton discusses in relation to digital food, '[i]n the context of the digital data economy, digitised information about food- and eating-related habits and practices are now accorded commercial, managerial, research, political and government as well as private value' (forthcoming: 11).

The Big Data economy – an economy marked by a growing IT investment in mapping behaviour trends and interactions through analysing very large data sets – is a space in which anyone can play if they have the right tools. It is the large global commercial food corporations who are, alongside government, best placed in terms of resources and access to use large-scale data monitoring for their own purposes, foregrounding what Mark Andrejevic calls the 'big data divide' (2014). As Stevens, Aarts, Termeer and Dewulf note in an essay on social media and agro-food:

> The food and beverage industry is at the forefront of interactive marketing and new types of digital targeting and tracking techniques. Food retailers

have taken over social media marketing companies to gain more data and enhance their marketing strategies.

(2016: 103)

Closely related to concerns over the commercialisation of personal data is the rise of what's been termed 'dataveillance', that is, the use of digital forms of connectivity to monitor individuals or groups (Clarke, 1988; van Dijck, 2014). Yet it is not only corporations who capitalise on such forms of surveillance. Related technologies of traceability also may enable activist groups and consumers to follow the complex globalised food chains of big agri-business and farmers to electronically monitor livestock and help prevent the spread of disease (Godsiff, 2016; Grindlay, 2016).

Such developments represent a double-edged sword for the realm of alternative food politics. Jaz Hee-jeong Choi and Mark Graham (2014), on the one hand, note the potential benefits that might arise from such food data. For instance, innovations like edible QR codes to tell customers where their dish has come from and apps that analyse photographs of food taken with the user's mobile phone to algorithmically identify its caloric content, 'have provided novel and richer ways for consumers to understand food' (ibid.: 152). Meanwhile, in the realm of agri-business, the rise of 'smart' farming is seeing the use of geo-tagged animals, data-driven production techniques and the use of drones to monitor crops and animals over large distances (Grindlay, 2016; Logan, 2017). Digital connectivity here is about the rationalisation and efficiency of systems and the maximisation of profit – with potential (though often short-term) benefits in terms of food production and food 'security' – but often at the cost of environmental and ethical scrutiny and sustainable and equitable approaches to land management.

The question of environmental cost, however, leads to a key but often ignored issue that is crucial for food activists, consumers, information technologists, supporters of e-democracy and corporates alike, that is, the negative environmental consequences of a growing reliance on technology. In an article for the online media studies forum *Flow*, Richard Maxwell and Toby Miller (2008) ask whether e-waste is 'the elephant in media studies' living room'. Proceeding to list the major environmental problems associated with a growing reliance on digital technologies in everyday life, from the toxic pollution produced from today's electronic media, for instance, the cadmium contained in mobile phones, and the growing impact of e-waste, much of which is exported to the Global South,[6] to the large amount of energy required both to produce digital technologies and associated infrastructure and to maintain and use the said technologies, they conclude with the following call to arms: 'We urge media scholars to take up the challenge of media technology's impact on the environment: recycle and rethink the life cycle of media technology within its ecological context, from design to disposal' (ibid.).

Likewise, for alternative food movements and environmental activists alike, it is important, as Sui and Rejeski (2002: 155) put it, 'not to treat the Internet as the Holy Grail for environmental salvation'. While there is an expanding critical literature on the environmental, resource, social and health implications of the growing reliance of environmental activists and conservationists on digital technology (Berkhout and Hertin, 2004; Arts *et al.*, 2015; Parikka, 2015; Rudram *et al.*, 2016), such debates can easily be pushed into the background by discourses of techno-optimism. An important task for food and sustainability activists engaging with digital technologies, then, is not only to address the costs and consequences of this, but to begin to imagine and design food futures that take the technological and infrastructural underpinnings of digital activism into account and that aim to develop more sustainable and equitable engagements with the digital realm.

Conclusion: governing and decoding digital foodscapes

This chapter has discussed the ways in which digital connectivity has become an important component of contemporary food politics and activism, from enabling otherwise disconnected households to support the broader urban farming movement through web-based communities such as permablitz through to enabling local food groups such as the Oklahoma Food Cooperative to share insights and resources with similar groups across the world. For consumers and households concerned with connecting with other like-minded food citizens, Web 2.0 has offered a cheap, accessible means to mobilise, link and collectivise. For alternative food producers and food communities, new media platforms have provided the communicative and connective infrastructure for re-scaling 'local' food hubs to offer real alternatives to global agri-business. As Stevens *et al.* put it:

> Flexible networks enable communities to join for a common cause in opposition against industrial food production, the horizontal links can bypass industrial-economic institutions and the interpersonal communication supports the social connections important to alternative food networks. This provides an opportunity for actors that do not fit industrial production standards, in which farmers play a greater role and the social and environmental origins of food products are emphasised.
>
> (2016: 103)

However, as this chapter has emphasised, while the rise of Web 2.0 may have initially offered seemingly 'neutral' spaces for the sharing of information and communication, the programming, formalisation and customisation of social media along commercial lines mean that many forms of online connectivity have increasingly become sites for capital to extract value (van Dijck, 2013). Another limitation with relying on online forms of connective action in the food realm is the fact that not all people are equally placed to engage with or have access to the connective realm (Dimon, 2013). For food activists, such a realisation requires maintaining a healthy scepticism towards the progressive participatory promise of the media tools and infrastructure that are increasingly dominating our daily lives.

This brings us to questions of visibility and transparency on the Internet. A recurrent trope in food activism oriented towards consumers is a concern with making visible the origins and conditions of global food production, often through the provision of information provided via apps or interactive online platforms. Political engagement and empowerment are thus equated with informational transparency (as per the Right to Know GMO movement[7]) that is, with the idea that by making visible some of the questionable production processes adopted by big food players so that consumers can choose *not* to buy their products, we can also challenge these increasingly normative food systems. As I have suggested, however, enabling consumers to choose alternative products may lead to political quietism, to a sense that the consumer act itself displaces the need for more collective forms of activism.

The question of informational transparency also raises issues of what we mean by 'transparency', and who has and does not have access to key forms of knowledge. As Arthur Mol argues, while the possession of environmental information in a digital era may enable some degree of democratisation of systems of governance and control, the rise of what he terms 'a new informational mode of environmental governance' also raises critical concerns about the potential for 'new power constellations' (2006: 511) around information access and use. As I have noted in this chapter, the digitisation of food through commercial apps like whatscook, for instance, can change our relationship to food in ways that are often *hidden* from view to the average person, embedded in an increasingly invisible algorithmic logic and culture.

In relation to food politics and questions of activism, the vocabulary of visibility, transparency and connection needs to be tempered by an awareness of issues of information governance and control. This, in turn, suggests that alternative food movements will need increasingly to understand and critically debate the social, cultural and political economies of digital data processes and infrastructures. As critical technology gurus Arthur and Marilouise Kroker put it:

> Technological society is no longer understandable simply in terms of the globalizing spectacle of electronic images but in the more invisible, pervasive, and embodied language of computer codes ... when codework becomes the culture within which we thrive, then we must become fully aware of the invisible apparatus that supports the order of communications within which we live.
>
> (2013: 7)

In a digital age, food citizens increasingly need to develop critical media literacy as much as food literacy. Just as the savvy food consumer is increasingly aware of the material conditions under which food is produced, so too online literacy requires a similarly critical approach to the material or infrastructural underpinnings of our everyday digital engagements. While digital connectivity is an increasingly taken-for-granted and ubiquitous part of political engagement, we do not necessarily stop to reflect on the material infrastructure required to support an increasingly globalised Internet (Horst, 2013). As with the realm of coded infrastructure, as I have argued above, the material foundations of the Internet also raise a larger set of political economy and governmental issues, as well as concerns about the potential environmental and social impact of digital infrastructure and energy-reliant communicative systems. Just as the food politics movement is seeking to build alternative sustainable ways of connecting all members of the food community, so too a digital citizenry needs to support systems that take into account questions of e-waste and sustainable physical and digital infrastructure design (Maxwell and Miller, 2011), and, as argued in this chapter, to work to foreground the social, political, ethical and material elements that underpin and constitute our digital pathways.

Acknowledgements

The author would like to thank Allister Hill and Jacinthe Flore for their excellent research assistant work and the editors of this collection for their extensive and very helpful critical feedback on an earlier draft of this chapter.

Notes

1 www.followthethings.com/grocery.shtml (accessed 20 March 2016).
2 http://oklahomafood.coop (accessed 24 June 2017).
3 A term first invented in Melbourne, a permablitz is an event where a group of volunteers come together to transform a backyard into a food-producing garden drawing upon the principles of permaculture.
4 The idea of permaculture was developed in the mid-1970s by Australians Bill Mollison and David Holmgren as an alternative to industrialised forms of agriculture. Conceived of as an ethical and holistic design system for sustainable living, land use and land repair, '[p]ermaculture has come to mean a design system, for taking patterns and relationships observed in natural ecosystems into novel productive systems for meeting human needs' and has been embraced by individuals, groups and communities worldwide (www.permablitz.net/resources/our-principles) (accessed 15 October 2017).
5 Monsanto's website offers a 'good news' message for the future, see https://monsanto.com (accessed 24 June 2017).
6 For instance, Caravanos *et al.*'s (2014) investigation of the impact of toxic waste on disability-adjusted life-years (DALYs), such as Chromium VI in India, indicates a higher burden of disease expressed in DALYs than other public health threats.
7 www.facebook.com/RightToKnowDe (accessed 24 June 2017).

References

Andrejevic, M. (2007) *ISpy: Surveillance and Power in the Interactive Era.* Kansas: University of Kansas.
Andrejevic, M. (2014) The big data divide. *International Journal of Communications*, 8: 1673–1689.
Amin, A. (2002) Spatialities of globalisation. *Environment and Planning A*, 34(3): 385–399.
Arts, K., Van der Wal, R. and Adams, W. M. (2015) Digital technology and the conservation of nature. *Ambio: A Journal of the Human Environment*, 44(Suppl. 4): 661–673.

Bennet, D. (2014) Inside Monsanto, America's third-most-hated company. *Bloomberg* [online], 4 July. Available at: www.bloomberg.com/news/articles/2014-07-03/gmo-factory-monsantos-high-tech-plans-to-feed-the-world (accessed 22 September 2016).

Bennett, L. W. and Segerberg, A. (2013) *The Logic of Connective Action: Digital Media and the Personalization of Contentious Politics.* Cambridge: Cambridge University Press.

Berkhout, F. and Hertin, J. (2004) De-materialising and re-materialising: Digital technologies and the environment. *Futures*, 36: 903–920.

Berlant, L. (1997) *The Queen of America Goes to Washington City: Essays on Sex and Citizenship.* Durham, NC: Duke University Press.

Booth, S. and Coveney, J. (2015) *Food Democracy: From Consumer to Food Citizen.* Dordrecht: Springer.

Bos, E. and Owen, L. (2016) Virtual reconnection: The online spaces of alternative food networks in England. *Journal of Rural Studies*, 45: 1–14.

Burgess, J. and Green, J. (2009) *YouTube: Online Video, and Participatory Culture.* Cambridge: Polity.

Caravanos, J. Gutierrez, L. H., Ericson, B. and Fuller, R. (2014) A comparison of burden of disease from toxic waste sites with other recognized public health threats in India, Indonesia and the Philippines. *Journal of Health and Pollution*, 4(7): 2–13.

Choi, J. H. and Graham, M. (2014) Urban food futures: ICTs and opportunities. *Futures: The Journal of Policy, Planning and Future Studies*, 62(Part B): 151–154.

Clarke, R. A. (1988) Information technology and dataveillance. *Communications of the ACM*, 31: 498–511.

Coff, C., Barling, D., Korthals, M. and Nielsen, T. (eds) (2006) *Ethical Traceability and Communicating Food.* Berlin: Springer.

Couldry, N. (2015a) Illusions of immediacy: Rediscovering Hall's early work on media. *Media, Culture & Society*, 37(4): 637–644.

Couldry, N. (2015b) The myth of 'us': Digital networks, political change and the production of collectivity. *Information, Communication & Society*, 18(6): 608–626.

Crang, P. and Cook, I. (2004) The world on a plate: Culinary culture, displacement and geographical knowledges. In: N. Thrift and S. Whatmore (eds) *Cultural Geography: Critical Concepts in the Social Sciences.* London: Sage.

Dimon, L. (2013) This is how disconnected the world is in the Internet age. *Mic*, 3 November. Available at: https://mic.com/articles/71111/this-is-how-disconnected-the-world-is-in-the-internet-age#.aXMnFyE3T (accessed 22 September 2016).

Eli, K., Dolan, C., Schneider, T. and Ulijaszek, S. (2016) Mobile activism, material imaginings, and the ethics of the edible: Framing political engagement through the Buycott app. *Geoforum*, 74: 63–73.

Foley, J. (2017) A five-step plan to feed the world. *National Geographic Magazine.* Available at www.nationalgeographic.com/foodfeatures/feeding-9-billion/ (accessed 22 September 2016).

Fuchs, C. (2013) Social media and capitalism. In T. Olsson (ed.) *Producing the Internet: Critical Perspectives of Social Media.* Göteborg: Nordicom, pp. 25–44.

Giddens, A. (1991) *Modernity and Self-Identity: Self and Society in the Late Modern Age.* Cambridge: Polity.

Godsiff, P. (2016) Blockchains could help restore trust in the food we choose to eat. *The Conversation.* Available from: https://theconversation.com/blockchains-could-help-restore-trust-in-the-food-we-choose-to-eat-62276 (accessed 26 April 2017).

Gómez-Benito, C. and Lozano, C. (2014) Constructing food citizenship: Theoretical premise and social practices. *Italian Sociological Review*, 4(2): 135–156.

Goodman, M. K., Maye, D. and Holloway, L. (2010) Ethical foodscapes?: Premises, promises and possibilities. *Environment and Planning A*, 42(8): 1782–1796.

Grindlay D. (2016) Victorian Government mandates electronic ear tags for sheep and goats. *The Conversation*. Available at: www.abc.net.au/news/rural/2016-08-24/sheep-electronic-tags-lambs-wool-policy-victoria-biosecurity/7779328 (accessed 26 April 2017).

Hatfield, L., *et al.* (2008) *Cultivating the Web: High Tech Tools for the Sustainable Food Movement*. New York: Eat Well Guide.

Hearn, G. *et al.* (2014) Using communicative ecology theory to scope the emerging role of social media in the evolution of urban food systems. *Futures*, 62(Part B): 202–212.

Holmgren, D. (2002) *Permaculture: Principles and Pathways Beyond Sustainability*. Hepburn: Holmgren Design Services.

Horst, H. A. (2013) The infrastructures of mobile media: Towards a future research agenda. *Mobile Media & Communication*, 1(1): 147–152.

Humphery, K. and Jordan, T. (2016) Mobile moralities: Ethical consumption in the digital realm. *Journal of Consumer Culture*. DOI: 10.1177/1469540516684188.

Kroker, A. and Kroker, M. (2013) *Critical Digital Studies: A Reader*. 2nd edn. Toronto: University of Toronto Press.

Leach, M., Scoones, I. and Stirling, A. (2010) *Dynamic Sustainabilities: Technology, Environment and Social Justice*. London: Earthscan.

Lewis, T. (2008) *Smart Living: Lifestyle Media and Popular Expertise*. New York: Peter Lang.

Lewis, T. (2012) 'There grows the neighbourhood': Green citizenship, creativity and life politics on eco-TV. *International Journal of Cultural Studies*, 15(3): 315–326.

Lewis, T. (2015) 'One city block at a time': Researching and cultivating green transformations. *International Journal of Cultural Studies*, 18(3): 347–363.

Lewis, T. and Huber, A. (2015) A revolution in an eggcup? Supermarket wars, celebrity chefs, and ethical consumption. *Food, Culture and Society*, 18(2): 289–308.

Lewis, T. and Potter, E. (2011) Introducing ethical consumption. In T. Lewis and E. Potter (eds) *Ethical Consumption: A Critical Introduction*. London: Routledge, pp. 12–44.

Local Dirt. (2016. Local dirt: Buy, sell and find local food. *Local Dirt*. Available at: http://localdirt.com (accessed 22 September 2016).

Logan, T. (2017) Drone mapping in agriculture on the rise. *ABC News*. Available at: www.abc.net.au/news/rural/2017-03-07/drone-use-increasing-for-ndvi-mapping/8328456 (accessed 26 April 2017).

Lupton, D. (2015) *Digital Sociology*. London: Routledge.

Lupton, D. (forthcoming) Cooking, eating, uploading: digital food cultures. In K. Le Besco and P. Naccarato (eds) *The Handbook of Food and Popular Culture*. London: Bloomsbury. Available at: http://papers.ssrn.com/sol3/papers.cfm?abstract_id=2818886 (accessed 22 September 2016).

Massey, D. (2004) Geographies of responsibility. *Geografiska Annaler: Series B, Human Geography*, 86(1): 5–18.

Maxwell, R. and Miller, T. (2008) E-waste: elephant in the living room. *Flow*, 2 December. Available at: www.flowjournal.org/2008/12/e-waste-elephant-in-the-living-room-richard-maxwell-queens-college-cuny-toby-miller-uc-riverside/ (accessed 26 April 2017).

Maxwell, R. and Miller, T. (2011) Eco-ethical electronic consumption in the smart design economy. In T. Lewis and E. Potter (eds) *Ethical Consumption: A Critical Introduction*. London: Routledge, pp. 141–155.

Micheletti, M. (2003) *Political Virtue and Shopping: Individuals, Consumerism, and Collective Action*. New York: Palgrave Macmillan.

Mol, A. P. J. (2006) Environmental governance in the information age: The emergence of informational governance. *Environment and Planning C: Government and Policy*, 24(4): 497–514.

Mollison, B. (1988) *Permaculture: A Designer's Manual*. Tyalgum: Tagari Publications.

Mollison, B. and Holmgren, D. (1978) *Permaculture One: A Perennial Agriculture for Human Settlements*. Tyalgum: Tagari Publications.

Monbiot, G. (2007) Environmental feedback: A reply to Clive Hamilton. *New Left Review*, 45: 105–113.

Monsanto (2015) A sustainable agriculture company. Available at: www.monsanto.com/pages/default.aspx (accessed 22 September 2016).

Oklahoma Food Cooperative. (2016) Available at: http://oklahomafood.coop/ (accessed 22 September 2016).

OLIO (2016) About. Available at: https://olioex.com/about (accessed 22 September 2016).

Open Food Network. (2015) Open Food Network. Available at: https://openfoodnetwork.org/ (accessed 22 September 2016).

Papacharissi, Z. A. (2009) The virtual sphere 2.0: The Internet, the public sphere and beyond. In A. Chadwick and P. N. Howard (eds) *The Routledge Handbook of Internet Politics*. London: Routledge, pp. 230–245.

Papacharissi, Z. A. (2010) *A Private Sphere: Democracy in a Digital Age*. Cambridge: Polity Press.

Parikka, J. (2015) *A Geology of Media*. Minneapolis, MN: University of Minnesota Press.

Peekhaus, W. (2010) Monsanto discovers social media. *International Journal of Communication*, 4: 955–976.

Pine, J. and Gilmore, J. (1999) *The Experience Economy*. Boston: Harvard Business School Press.

Rudram, B., *et al.* (2016) The impact of digital technology on environmental sustainability and resilience: An evidence review. *Institute of Development Studies*. Available at: www.ids.ac.uk/publication/the-impact-of-digital-technology-on-environmental-sustainability-and-resilience-an-evidence-review (accessed 26 April 2017).

Schor, J. (1999) The new politics of consumption. *Boston Review*. Available at: http://new.bostonreview.net/BR24.3/schor.html (accessed 22 September 2016).

Stevens, T. M., Aarts, N., Termeer, C. J. A. M. and Dewulf, A. (2016) Social media as a new playing field for the governance of agro-food sustainability. *Current Opinion in Environmental Sustainability*, 18: 99–106.

Suarez-Villa, L. (2016) *Globalization and Technocapitalism: The Political Economy of Corporate Power and Technological Domination*. Farnham: Ashgate.

Sui, D.Z. and Rejeski, D. W. (2002) Environmental impacts of the emerging digital economy: The e-for-environment e-commerce? *Environmental Management*, 29: 155–163.

Terranova, T. (2004) *Network Culture: Politics for the Information Age*. London: Pluto Press.

Van Dijck, J. (2013) *The Culture of Connectivity: A Critical History of Social Media*. Oxford: Oxford University Press.

Van Dijck, J. (2014) Datafication, dataism and dataveillance: Big data between scientific paradigm and ideology. *Surveillance & Society*, 12(2): 197–208.

Viswanathan, S. (2010) Inside the mind of George Consagra. *Good Guide Blog*, 16 September. Available from: https://blog.goodguide.com/2010/09/16/inside-the-mind-of-george-consagra/?like=1 (accessed 22 September 2016).

Wilkins, J. L. (2005) Eating right here: Moving from consumer to food citizen. *Agriculture and Human Values*, 22(3): 269–273.

World Economic Forum (2016) Are you feeling it? Why consumer companies must master the experience economy. *World Economic Forum*. Available at: http://reports.weforum.org/digital-transformation/moving-to-the-next-level-the-experience-economy/ (accessed 26 April 2017).

11 Digital food activism

Values, expertise and modes of action

*Karin Eli, Tanja Schneider, Catherine Dolan
and Stanley Ulijaszek*

Introduction

In this chapter, we turn to the three case studies we explored in our 2013–2016 research project on how new information and communication technologies (ICTs) mediate novel forms of consumer activism and food governance. Each case study represents a different type of digital platform used in food activism: a mobile app, a wiki platform, and an online-centric activist organization. Case studies, by definition, do not provide an exhaustive review; for example, the case studies we selected do not address more diffuse activism through social media. Our aim, however, is not to review all types of digital food activism, but rather to capture diverse forms and potentials of digital food activism, and develop an analytic framework that can be applied to other cases in the field.

What do we mean by digital food activism? As we define it, digital food activism does not simply refer to food activism that occurs on digital media. Rather, it encompasses forms of food activism enabled and shaped by and through digital media platforms. Traditional food activism – for example, the work of alternative food networks (AFNs), such as Fair Trade, which focus on developing 'economic and cultural spaces' for the ethical production, distribution, and consumption of foods (Goodman and Goodman, 2009: 209) – also employs supportive digital elements, including websites, blogs, and social media presence to enhance their consumer base (Lekakis, 2014). Digital food activism, by contrast, occurs largely on and through digital media platforms, with the medium as a central part of the message. Digital platforms, such as social media, are then used to reimagine these traditional forms of AFN activism, creating new messages and activist publics – and, in the case of Fair Trade, implicating producers rather than consumers (see Lyon, Chapter 4 in this volume). The contrast between activism that occurs on digital media and activism shaped by digital media echoes Vegh's (2003) distinction between 'Internet-enhanced' and 'Internet-based' activism.

In each of our case studies, the origins, development, and implementation of the activist project were interwoven with a digital platform. These digital platforms, then, were conceptualized by the social entrepreneurs who developed them not as supporting consumer action, but as fostering and mediating both consumer and producer activism. The activism they facilitate, therefore, is 'Internet-based', rather than 'Internet-enhanced' (ibid.).

In comparing the forms of 'Internet-based' activism facilitated in the case studies, we therefore attend to the digital platforms that underlie them. 'Digital devices', as Ruppert *et al.* (2013) argue, 'are simultaneously shaped by social worlds, and can in turn become agents that shape those worlds'. This co-constitutive shaping develops through the dynamic creation of logic, involving both humans and devices. According to Van Dijck and Poell (2013), the 'logic' of social media, while initially created by developers, is then 'distribute[d]' by platforms, adapted by users, integrated by institutions, and continuously updated by developers to influence and respond to user and industry dynamics. Through a complex of 'sociotechnical processes', human action and connection begin to incorporate social media logics. For example, in Bucher's (2013) analysis of Facebook, she finds that the concept and actualization of friendship on the platform inherently involve nonhuman actors and logics that become seamlessly integrated into the 'making' of human connections. Thus, to follow Sayes (2014), the digital platform is a mediator that continuously alters human relations, dynamically connecting human actors through non-human spaces and processes.

Drawing on actor-network theory, we suggest that digital platforms are not merely utilitarian objects, secondary to the principles of food activism. Previous research in sociology, geography, and critical food studies has challenged the imagined divisions between human and non-human in food networks, calling for a consideration of animals, crops, food products, and agricultural technologies as food network actors (Lockie, 2006; Morris and Kirwan, 2011; Roe, 2006). We extend these critiques to the activist realm, and suggest that digital technologies play a formative role in emergent forms of food activism. Through a comparative analysis of three case studies, we pose a set of core questions: (1) who sets the agenda, and what values guide this agenda?; (2) what evidence counts, and who has the expertise to provide it?; and (3) what modes of action are employed, and through which constituencies (i.e. groups of supporters, users, or consumers)? Our answers suggest that it is in the interaction of activist entrepreneurs, consumer-citizens, producers, retailers, and digital platforms that new forms of food activism emerge.

Methods

Our analysis focuses on case studies of three activist organizations. We adapted our methods to each case study to reflect the types of organizations involved, the forms of data they produced, and the scale of their activist projects, as well as the level of access we could have to the workings of each.

Our mobile app case study focuses on Buycott. Buycott is a US-based barcode scanner app, with a global database that encompasses a range of retail products, including, predominantly, food. Buycott's stated mission is to enable consumers to 'vote with [their] wallet[s]' (Buycott, 2016). Unlike other barcode scanner apps (such as Fooducate and the Good Guide), which offer authoritative information on the 'goodness' of food content – both nutritional and ethical – Buycott engages in a different type of knowledge production, one that positions the app itself as a neutral platform. App users generate activist campaigns, and provide both data and judgement; other users join those campaigns. When a user scans a product with Buycott, the app alerts the user to conflicts between the campaigns the user has joined and the politics of the company behind the product (see also Eli *et al.*, 2016). In our analysis of Buycott, we focused on consumer, app developer, and news media framings of Buycott's two largest campaigns (ibid.). To obtain additional information about the operations of Buycott, Inc., the company behind the app, we conducted a brief Skype interview with company founder Ivan Pardo in February 2017.

Our wiki platform case study focuses on a Central European organization we call by the pseudonym HowToBuyWiki. HowToBuyWiki is a non-profit organization dedicated to the development of an emerging, open-source Internet platform which, as the members describe it, is designed to enable 'conscious consumption' through promoting 'product transparency'. With the wiki platform focused on the structuring and sharing of comparative information about products, as provided by wiki editors and users, HowToBuyWiki engages in a data-centric form of consciousness-raising among consumers. Founded in 2006 by a group of friends and registered as a charitable organization, HowToBuyWiki advocates a 'civil society' concept that emphasizes their organization's independence from governmental, commercial, and third-sector institutions. Our case study of HowToBuyWiki relied on thematic analysis of interviews and documents provided by the organization's members, alongside our own exploration of the wiki platform as (non-contributing) users. From May to October 2013, we interviewed all six active members of HowToBuyWiki. In addition, HowToBuyWiki's core members provided us with meeting minutes, the HowToBuyWiki constitution, and other related documents, which we have also incorporated into the analysis (see Eli *et al.*, 2015). Given the access we had to this small organization's internal memoranda and behind-the-scenes negotiations, we have chosen to use a pseudonym to protect the confidentiality of the members.

Our case study of an online-centric consumer advocacy organization focuses on foodwatch. foodwatch describes itself as 'an independent, non-profit organisation that exposes food-industry practices that are not in the interests of consumers' (foodwatch, 2016). Using a number of digital media tools, including e-newsletters, Twitter, and Facebook, foodwatch campaigns for a number of targeted causes by appealing for consumer support in the form of e-letter writing, donations, and membership. The main communication tools foodwatch employs are its website and its weekly e-newsletter, which is sent out to subscribers via email, and can either be read on or downloaded from the foodwatch website. Our case study is based on a thematic analysis of 50 e-newsletters that foodwatch published during 2013 (see Schneider *et al.*, under review).

The analysis that follows draws on our detailed investigation of these three case studies (Eli *et al.*, 2015; Eli *et al.*, 2016; Schneider *et al.*, under review), to identify key features that define digital food activism.

Who sets the agenda, and what values guide this agenda?

All three digital platforms represent the visions of activist entrepreneurs, who continue to lead their respective organizations. The organizations' agendas, however, are set by different actors. While Buycott developer Ivan Pardo initially branded the app as supporting particular agendas – namely, the labelling of genetically modified foods, and the avoidance of prominent Republican Party funders Koch Industries – he currently positions the app as a neutral space, to be shaped by app users. Accordingly, Buycott features multiple, and sometimes conflicting, activist agendas as defined by user-generated campaigns. Along similar lines, while HowToBuyWiki's members created their organization to support 'conscious consumption', with particular attention to sustainability and transparency, they developed their wiki platform as a neutral space, where agendas would be set by users (whom they call 'wiki editors'). In contrast, foodwatch, which is led by former Greenpeace Germany director Thilo Bode, is the only organization of the three that sets its agenda rather than acting as a platform for agendas set by users. Not incidentally, of the three, it is also the only organization that has a geographic presence, with offices in foodwatch's three sites – Berlin, Amsterdam and Paris.

While Buycott and HowToBuyWiki emphasize user-led agenda setting, the forms of agenda setting they enable are structured by the organizations' use of digital platforms for interacting with their users and each organization's core values. Buycott's core values are centred on promoting corporate transparency, exposing the political connections hidden in retail products, and enabling consumers to use the market as a political arena, with boycott and buycott action conceptualized as impacting on corporations and policy-makers. As such, the user-led agendas set on and through Buycott involve, by design, the revealing of corporate relationships and of connections between corporations and political causes. For example, Buycott's GMO labelling campaign highlights that, in the US, Monsanto has financed advertising to oppose propositions for the mandatory labelling of genetically modified foods.[1] When creating a campaign, users are prompted to enter the names of companies to oppose and companies to support, based on their connection to the cause; when users scan a product, a corporate kinship chart is generated to reveal connections between the product, its manufacturer, other companies, and parent companies. On HowToBuyWiki's platform, the organization's shaping of agenda-setting is equally pronounced. User-led agendas must follow HowToBuyWiki's inbuilt construction of product transparency as the organizing principle for data entry and presentation. HowToBuyWiki's core values are product transparency, exposing the attributes of products (that are often omitted on labels), enabling informed comparison within product categories, facilitating conscious consumption practices, and changing power relations in the market. In designing the wiki platform, HowToBuyWiki reflects these values through constructing wiki pages based on comparative tables divided by product category, wherein products are compared across several attributes; the attributes assigned to the chocolate table, for example, include brand, ingredients, allergens, and organic labelling, among others (see Eli *et al.*, 2015). In the case of foodwatch, the core values guiding the organization's agenda setting are food industry transparency, truth in advertising; focus on food safety, health and food security, and linking local and global food issues. Notably, unlike Buycott and HowToBuyWiki, foodwatch has core values that do not directly involve everyday consumer action in the form of consumption choices. Instead, these values lend themselves to foodwatch's approach to consumer action: calling for letter-writing campaigns.

As Table 11.1 shows, while all three organizations aim to promote transparency, their actions are directed at different aspects of the food landscape – corporations, consumer products, and the food industry. These differences in focus, then, link with the organizations' differing conceptualizations of who sets their agendas: whereas Buycott and HowToBuyWiki focus on action at the retail end and conceptualize consumers as setting the agenda, foodwatch focuses on action at the often-inaccessible macro-level of the food industry and positions itself as the agenda-setter.

What evidence counts, and who has the expertise to provide it?

Data are at the heart of all three activist platforms – but the types of evidence that count and the expertise that supports it are diverse. Buycott relies on data provided, mostly, by campaign organizers and subscribers. On the Buycott website, each campaign has a dedicated page, with a list of companies to be boycotted or buycotted; there, links are provided to support some claims, though the quality of those sources varies widely. Expertise is crowd-sourced; names of

Table 11.1 The organizations' agenda setting and values in comparison

	Buycott	*HowToBuyWiki*	*foodwatch*
Organizers	App developer Ivan Pardo	The HowToBuyWiki organization members	Thilo Bode and the foodwatch organization
Agenda	Set by campaign organizers (and moderated by Buycott) ('users')	Set by wiki editors ('users')	Set by foodwatch (not user-driven)[1]
Core values	Corporate transparency; exposing the political connections hidden in retail products; enabling consumers to use the market as a political arena	Product transparency; exposing the attributes of products (that are often left off labels) and enabling informed comparison within product categories; facilitating conscious consumption practices; changing power relations in the market	Food industry transparency; truth in advertising; focus on food safety, health and food security; linking local and global food issues

Note
1 No information on how campaign topics are chosen is available on the foodwatch website.

expert scientists or activists are not invoked, and the reliability of data and claims depends on the collective of Buycott users. Crowd-sourced data are also central to the HowToBuyWiki project, although, in practice, much of the data currently on the wiki platform was provided by the HowToBuyWiki members themselves, because the project had not reached critical mass at the time of writing. Evidence is dependent on corporations, as HowToBuyWiki data are gleaned from product labels, websites, and information provided by producers; HowToBuyWiki members emphasize that wiki users are free to debate the reliability of data on the wiki page of any product category, where multiple contesting sources may exist. HowToBuyWiki's project is supported by an expert advisory board, which includes journalists, consumer advocates, and policy-makers, whom HowToBuyWiki invited to consult and raise the profile of the organization (while leaving agenda-setting to wiki users); the advisory board, however, is not involved in data-related matters. Of the three organizations, foodwatch alone relies on its own expertise. foodwatch provides the data to support its campaigns, and invokes the organization's own status as evidence for the reliability of claims. Few references are made to named external experts, data are not crowd-sourced, and – unlike Buycott and HowToBuyWiki – foodwatch does not facilitate consumer feedback on or dialogue about the data provided, although followers of foodwatch on social media (e.g. on Facebook) occasionally discuss foodwatch campaigns on the organization's pages.

Table 11.2 shows the different forms of evidence and expertise that each organization emphasizes and claims. The crowd-sourced data models employed

Table 11.2 The organizations' forms of evidence and expertise in comparison

	Buycott	*HowToBuyWiki*	*foodwatch*
Data	Mostly provided by campaign organizers and subscribers	Crowdsourced, in principle (mostly provided by HowToBuyWiki, in practice)	Provided by foodwatch; mostly secondary data, e.g., published reports (although, in certain cases, foodwatch has funded research to obtain data).
Evidence	On the Buycott website, links are provided to support some claims implicating companies in campaigns	Product data are gleaned from labels, websites, and producer information; the organization suggests that sources be provided to support the information editors add to wiki pages; reliability can be debated on wiki pages	Claims rely on the organization's own expertise, research, and some publicly available documents.
Expertise	Crowdsourced	Crowdsourced; advisory board (journalists, consumer advocates, and policy-makers) to provide legitimacy	foodwatch

by both Buycott and HowToBuyWiki eschew a reliance on named, authoritative experts. Both organizations cultivate an image of collective grassroots action driven by pseudonymous concerned citizens, identified only by their screen-names, who pool their knowledge to mobilize for product and corporate trans-parency. While HowToBuyWiki has assembled an expert advisory board, it draws on their expertise to lend legitimacy to the project, but not to legitimize wiki editors' claims about products or industry. In contrast, the data model employed by foodwatch – in which data are presented as persuasive evidence to generate public support for the organization's campaigns – calls for the claiming of authoritative expertise. Thus, whereas Buycott and HowToBuyWiki create (the appearance of) a flattened hierarchy of political participation through crowd-sourcing data, foodwatch maintains, and indeed reinforces, hierarchies of polit-ical voice and expertise through presenting data to consumers, rather than involving consumers in data generation and discussion.

What modes of action are employed, and through which constituencies?

Each of the organizations envisions a different mode of consumer action. Buycott concentrates on users' everyday shopping practices. Guided by the motto 'voting with your wallet', the mode of action promoted on Buycott is con-sumer boycott or 'buycott' of retail products. As framed by Buycott, this action is generated by users, who set campaigns and provide data on companies, facilit-ated by the app. The active constituency for this action is comprised of Buycott users who subscribe to a given campaign; action is predicated on ownership of a smartphone or tablet, and on literacy with the use of mobile apps.

While HowToBuyWiki aims to foster 'conscious consumption', it envisions a mode of action that involves fostering discussion and corporate accountability through the wiki site. As the members of HowToBuyWiki frame it, their project's main goal is to change market dynamics, giving more weight to the voices of consumers; a wiki platform, they argue, provides a space where con-sumers can challenge corporations. The action they envision is facilitated by the platform and generated by wiki editors, who create wiki pages for products, with tables that compare the attributes of similar products from various brands. However, at the time of the study, the platform, which requires knowledge of wiki programming code, was difficult to edit. In response to an earlier draft of this manuscript, a leading member of HowToBuyWiki has informed us that, in the intervening years, the organization has taken steps to simplify the editing process, but that certain elements of the wiki editing processes, such as making product lists on the wiki platform, remain difficult. HowToBuyWiki's primary constituency is, therefore, wiki editors, although the organization also imagines the general public – or 'passive' readers – as acting on the information generated on and through the site.

Unlike the everyday consumption or dialogue between consumers and producers advocated by Buycott and HowToBuyWiki, foodwatch employs online petitions, email writing campaigns, pressure on public figures and decision-makers, and public shaming of companies as its mode of influence. This action is generated and directed by foodwatch, which mobilizes its active constituency of subscribers and the news media to facilitate its campaigns. Compared to the other two organizations, the action facilitated by foodwatch requires only basic technological infrastructure and literacy, as it is predicated on signing petitions via email or social media accounts. foodwatch's use of the Internet for campaigning, or 'e-mobilization', allows the organization to 'capitalize on its potential for recruitment, fund-raising, organizational flexibility and efficiency' (Chadwick, 2006: 115).

Modes of action are also influenced by legal status and funding; here, the three organizations show considerable diversity. Of the three organizations, Buycott, is the only one that was founded and is operating as a corporation. Ivan Pardo, the company's founder, explained that the decision to register Buycott, Inc. as a company rather than as a non-profit organization was due to the ease of incorporating as a company in California (pers. comm, 22 February 2017). Buycott's website provides little information on the company's core team and no information on its financial situation. At the time of writing, Internet searches for information on Buycott's stock volume, shareholders, annual report or funding sources have not yielded any data. Initial news media reportage on Buycott stated that Pardo, a digital entrepreneur, had originally bootstrapped[2] the development of the app and received help from two friends who 'pitched in to promote the app' (O'Connor, 2013), although their possible financial contribution was not mentioned. Since the app is free to download, no revenue is made from the sale of the app directly. However, the app now includes a revenue-generating upgrade: an in-app recommendation for alternative products, with links to online shops where these products can be purchased. While the app and the Buycott website are both updated regularly, it is unclear how many staff are involved in the technical and financial operations related to the app. In our interview, Pardo explained that his vision for Buycott focused on growing it as an activist platform that enables users to quickly discover relevant and actionable information. He emphasized that he was not motivated by achieving high financial returns on investment, and that he was content with Buycott generating just enough income for him 'to pay the bills', thereby allowing him to continue to develop the platform full-time. When asked about Buycott's investors' expectations, Pardo admitted that some would prefer the company to generate more profit, but that most investors were relaxed about revenue as their investments were ideologically, rather than financially, motivated.

In contrast, HowToBuyWiki, a registered society in its country of origin and, as such, a non-profit organization, includes financial information on its publicly accessed wiki platform. HowToBuyWiki explains in one sub-section of the wiki platform that membership fees and donations assist with financing the costs associated with server rental and, ideally, could be used for future employment of administrative staff tasked to run the platform. Indeed, currently, HowToBuyWiki's main way of sustaining the platform and the group's activities financially are membership fees paid by the HowToBuyWiki organization members and donations. From our conversations with HowToBuyWiki's members and analysis of the documents they shared with us, we know that the organization has applied for funding from independent private foundations that support civil society efforts and visions; to date, these applications have been unsuccessful. The lack of funding has slowed the project's progress, as the organization's members are only able to develop the platform on an uncompensated, part-time basis, with no funds to hire external support.

foodwatch and its team of 13 full-time and 5 part-time members are financed by support on a much larger scale, primarily through membership fees and donations. In comparison to Buycott, foodwatch, which is registered in Germany as a charitable organization, provides detailed information about the organization's finances on its website. On the foodwatch German language website, clicking on 'About foodwatch' and then on the subsection submenu 'Finances and Transparency'[3] leads to a page that describes foodwatch as striving for independence, and explains that foodwatch accepts neither state funding nor donations by the food industry or larger companies in the food retail sector. To achieve this, foodwatch states that it checks whether donors of sums greater than 500 euro have any ties to these sectors. Moreover, donors of sums greater than 5,000 euro are listed by name on the organization's website. foodwatch also provides details of its earnings and expenditures on its website (the latest available are for 2014, with comparisons to 2012 and 2013). foodwatch's financial transparency echoes its broader values and aims to achieve more transparency in the food industry. The organization is part of 'Initiative Transparente Zivilgesellschaft' (Transparent Civil Society Initiative) organized by Transparency International in Germany that stipulates that each civil society organization should disclose ten items/facts such as its constitution, the names of its key decision-makers, and personnel structure, as well as information about sources and application of funds to the public.[4] foodwatch adheres to these requirements and, in addition, offers information about the organization's financial funding process as a 'start-up-NGO' in 2002.

While both Buycott and foodwatch use campaigning as their mode of action, as Table 11.3 shows, the campaigning these organizations facilitate takes either a grassroots or a top-down approach, in concert with each organization's framing of constituency and organizational identity (as for-profit or charity). Structured as a 'traditional' non-profit, foodwatch has the organizational structure and the funding to operate as an expert organization that generates campaigns. By contrast, Buycott's reliance on user-generated data dovetails with its incorporated status, which opens its potential for third-party revenue, for example, via the distribution of user reviews to social media sites (Buycott, 2015) and by referring users to online stores where 'alternative' products can be purchased (pers. comm., 22 February 2017). In our view, this highlights the extent to which some forms of digital food activism have the potential to become an economic activity for those who set up, develop, and own platform and data. HowToBuyWiki, on the other hand, conceptualizes data as generated by wiki editors rather than by individually identifiable users, and, as a registered charity, cannot employ user-generated data to grow the company.

Table 11.3 The organizations' modes of action and active constituencies in comparison

	Buycott	HowToBuyWiki	foodwatch
Mode of action	'Voting with your wallet': consumer boycott/buycott of retail products; action generated by users and facilitated by app	Fostering discussion and corporate accountability through wiki site; action generated by users and facilitated by wiki	Online petitions, email writing campaigns and public shaming of companies; action generated by foodwatch, which mobilizes subscribers and media
Active constituency	Campaign creators and subscribers; subscriptions are linked to users' email, Twitter, or Facebook accounts (largest campaign as of 13 February 2017: 'Pro-GMO or Pro-Right to Know?'; 523,193 subscribers)[1]	Wiki editors, as well as the general public (wiki readers)	foodwatch members, newsletter subscribers, and Facebook followers (number of 'likes' on official Facebook page, as of 13 February 2017: 363,893)[2]
Legal status	Incorporated, domestic stock (California, USA)	Registered as a charitable organization (in country of origin)	Registered as a charitable organization (Germany)
Financing	Bootstrapping before app launch; supported by a number of investors[3]	Membership fees paid by the core committee; donations from family and friends; (unsuccessful) applications to civil society organizations	Membership fees and private donations

Notes
1 https://buycott.com/campaign/211/pro-gmo-or-pro-right-to-know (accessed 13 February 2017).
2 www.facebook.com/foodwatch/ (accessed 13 February 2017).
3 This information is based on personal communication with- Ivan Pardo, Buycott's founder (22 February 2017).

Discussion

While digital food activism has a lot in common with alternative food networks – in foci and in concepts of activism as fostered through 'conscious' consumption and non-consumption – it weaves together digital platforms with activist values and modes of action, so that technology and values are bound together. In each of the case studies, the activist systems involve similar actors, yet the vectors of action and influence are different. These vectors reflect not only the core values of each activist organization, but also an imagining and interpretation of values and preferred action that are closely connected with each organization's digital platform. The comparison of case studies therefore reveals the plurality of digital food activism – a plurality closely linked to the platforms used. The comparison also reveals the limits of these platforms, whose use is always-already constrained by technological literacy and infrastructure.

In the logic of interaction configured by Buycott (Figure 11.1), the app itself – and the organization that developed it – are positioned as mediators. While the organization itself mobilizes media support and empowers consumers, it does not directly leverage this mobilization to influence corporations and policymakers. Rather, as members of the general public become Buycott subscribers, and as subscribers reinforce or start campaigns on the mobile app, they act through Buycott to mobilize for political change, with performances of product boycotts and buycotts online (and presumably also offline).

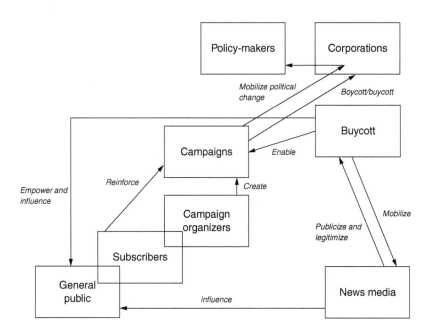

Figure 11.1 Buycott's logic of interaction.

Similarly, in the logic of interaction envisioned by HowToBuyWiki (Figure 11.2), as members of the general public become 'conscious consumers', that is, wiki editors or readers, they are able to use the wiki site to pressure corporations (or as HowToBuyWiki phrases it, producers). However, in constructing a fairly difficult-to-edit wiki platform, HowToBuyWiki also configured a more central role for itself, with members of the organization acting as wiki editors. In addition, unlike Buycott, which builds its legitimacy through news media coverage and numbers of subscribers, HowToBuyWiki seeks to legitimize itself as an organization through the more traditional means of convening an expert advisory board, reinforcing the centrality of the organization's core membership in defining and sustaining its project.

Unlike Buycott and HowToBuyWiki, foodwatch places itself as the main agent in its logic of interaction (Figure 11.3). The organization mobilizes subscribers and news media alike to legitimize its project; through media coverage and numbers of subscribers, foodwatch gains the necessary authority to protest and expose corporations, and petition and accuse government ministers of collusion with a food industry that does not respect consumer interests and human welfare. Although foodwatch positions these actions as representing the interests of the general public, consumers act through foodwatch only in adding their names to pre-written campaigns, and do not participate in its agenda setting; rather, their role is to empower this central organization as a sole agent.

While food activism, as defined by Siniscalchi and Counihan (2014: 3), refers to 'efforts by people to change the food system across the globe by modifying how they produce, distribute, and/or consume food', our concept of digital

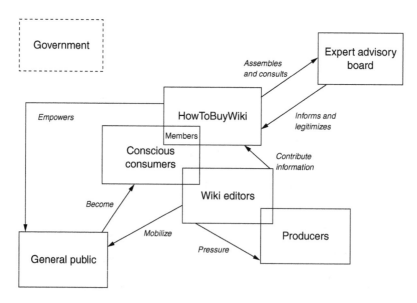

Figure 11.2 HowToBuyWiki's logic of interaction.

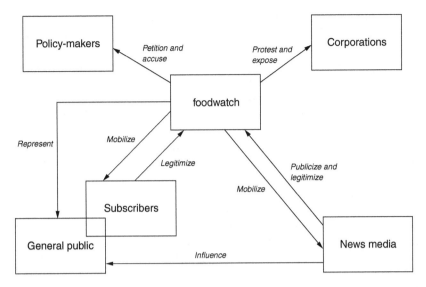

Figure 11.3 foodwatch's logic of interaction.

food activism extends beyond immediate concerns with the food system, to encompass activist projects where food is a means to broader political action. Based on our case studies, we suggest that at the core of digital food activism is the changing of market relations. Siniscalchi and Counihan observe that

> [F]ood activists often try to exploit the community dimension [of the economy] to affect the market dimension, for example, by creating farmer's markets or fair trade regimes to link producers and consumers in personal relationships that accrue economic benefits to both.
>
> (Ibid.: 10)

All three organizations appeal to this 'community dimension'. Buycott and How-ToBuyWiki encourage users to upload information about products and companies for the benefit of other users, while foodwatch mobilizes consumers to share foodwatch's insights with members of their social networks, and to partake in collective email petition writing. In all three cases the community dimension is used to develop consumer publics that form around products and issues, with the aim of achieving greater transparency, fairness, and justice through getting closer to food's origin (cf. Schneider *et al.*, under review). Entangled in these ambitions is a commitment to affect markets. By shifting flows of information, Buycott and HowToBuyWiki seek to shift and redirect flows of money to alternative food (and other product) networks. In the case of foodwatch, providing and promoting alternative information on products and companies is envisioned as affecting the governance of markets and marketing.

In developing food issue publics, these three organizations also generate activist or business models that could, in theory, allow for growth. While the participatory 'community dimension' is central to the missions of these organizations, community formation can also be viewed as a means of 'outsourcing' costs through increasing the numbers of users who contribute to the platform's content or act as lay ambassadors of the organization, thus providing their knowledge, expertise, or marketing support for free. This configuration positions consumers not simply as activists but as producers – or 'prosumers'[5] (Gabriel and Lang, 2015) – a process characteristic of the sharing and platform economy.[6]

While Buycott, HowToBuyWiki, and foodwatch all privilege transparency as an organizing value, we found that the organizations locate transparency in different nodes of the production chain. While foodwatch calls for transparency in marketing, production processes, and political lobbying, HowToBuyWiki emphasizes ingredients and environmental impact, and Buycott focuses on exposing corporate ownership. These definitions of transparency inform, and are also shaped by, the platforms themselves; for example, ingredients and sustainability feature centrally as attributes in HowToBuyWiki's comparative tables, while corporate kinship charts are key to Buycott's visualizing of campaigns. Thus, while certain aspects of a food product or its manufacturer are rendered visible through these platforms, other potential aspects remain more obscure. We suggest, therefore, that in the realms of food production, distribution, and consumption, the definition of transparency is wide-ranging and organizations aiming to make food-related processes more transparent make choices that (knowingly or unknowingly) contribute to qualifying what food transparency is. To make this organizational qualification process visible, the organizations we studied might regularly disclose how they decide to focus on certain aspects of food production, distribution, and consumption. HowToBuyWiki, for example, might promote further transparency on the wiki platform by stating on each wiki page (and encouraging wiki editors to do so, as well) why particular product categories and attributes were chosen. Likewise, foodwatch might offer information on how they select the food issues highlighted in their campaigns, among a range of possible topics. Buycott might explain why corporate ownership is key to transparency, where food-related issues and campaigns are concerned.[7]

To understand the implications of digital platforms, we must address their *modus operandi* – the productive logics that structure them, and that, in turn, enable them to structure human action (Ruppert *et al.*, 2013). Such 'reverse engineering', to use Fox's (2015) phrase, reveals the micropolitics and possibilities that inhere in digital platforms. In the realm of digitally enabled activism, other scholars have called attention to new forms and concepts of action generated by and through digital platforms – and, in particular, social media. In an analysis of activism on Facebook, Langlois *et al.* (2009) have suggested that 'online platforms' define new 'issue publics', through offering a means of association and action to individuals who would not have connected otherwise. Bennett and Segerberg (2012) have argued that digital platforms also provide a new activist logic, with digitally enabled activism based in 'connective', rather than collective action – a logic that transposes activism from centrally directed, organized action to diffuse, individual sharing of data.

As the analysis we have presented here has shown, differences between digital food activism platforms reflect different activist logics. These logics are, in turn, reproduced through the digital platforms themselves, and the potentialities for action that they offer, enable, and direct. The digital platforms, then, become part of the activist landscape, revealing and structuring diverse (and at times divergent) pathways to digital food activism.

Acknowledgements

We would like to thank HowToBuyWiki's members and Buycott Inc. founder Ivan Pardo for generously sharing information about their organizations. Thanks also go out to Amy K. McLennan and Javier Lezaun, who co-authored two of the case study papers that stemmed from this project. This chapter is based on research funded by the Oxford Martin Programme on the Future of Food, and we are thankful for the Oxford Martin School's generous support.

Notes

1 See http://buycott.com/campaign/companies/211/pro-gmo-or-pro-right-to-know (accessed 7 March 2017).
2 To bootstrap is a financial/investment term that specifies that entrepreneurs start a company by drawing on their personal finances or the operating revenues of the new company, rather than seeking a bank loan or other forms of investment.
3 For detailed information, see www.foodwatch.org/de/ueber-foodwatch/finanzen-transparenz/ (in German, accessed 2 August 2016).
4 For more information about foodwatch's participation in the initiative, see www.food watch.org/de/ueber-foodwatch/finanzen-transparenz/transparenz-initiative/ (in German, accessed 2 August 2016). For more information about the initiative for transparent civil society, see www.transparency.de/Initiative-Transparente-Zivilg.1612.0.html (in German, accessed 2 August 2016). Some information is accessible in English, too: www.transparency.de/English.1222.0.html (accessed 2 August 2016).
5 Gabriel and Lang (2015: xi) explain that '"prosumer"' is 'a word combining "producer" and "consumer" coined by futurologist Alvin Toffler to denote that the work process is being incorporated into the consumption process and vice versa'.
6 While all three organizations have potential for development through crowd-sourced consumer action, the growth trajectories they have followed diverge considerably. Reasons may include differences between the organizations' geographic foci, legal status, news media, and public engagement, and the accessibility of their user interfaces, among others.
7 Another aspect of transparency is organizational. In attending to the organizations' legal statuses and funding statements, it becomes apparent that the provision of such information, rather than being built into the platforms, is entirely dependent on the organizations' own disclosure statements. Of note, Buycott – the only organization among the three not to be registered as a charity – has not been challenged by users to disclose its financial information, despite the fact that the Buycott app targets user concerns over corporate transparency, suggesting that the concept of transparency may carry different value judgements when a digital, crowd-sourced activism platform is involved.

References

Bennett, W. L. and Segerberg, A. (2012) The logic of connective action: Digital media and the personalization of contentious politics. *Information, Communication & Society*, 15(5): 739–768.

Bucher, T. (2013) The friendship assemblage investigating programmed sociality on Facebook. *Television & New Media*, 14(6): 479–493.

Buycott. (2015) Terms and conditions. Available at: www.buycott.com/terms (accessed 14 February 2017).

Buycott. (2016) Vote with your wallet. Available at: www.buycott.com (accessed 8 August 2016).

Chadwick, A. (2006) *Internet Politics: States, Citizens, and New Communication Technologies*. Oxford: Oxford University Press.

Eli, K., Dolan, C., Schneider, T. and Ulijaszek, S. (2016) Mobile activism, material imaginings, and the ethics of the edible: Framing political engagement through the Buycott app. *Geoforum*, 74: 63–73.

Eli, K., McLennan, A. K. and Schneider, T. (2015) Configuring relations of care in an online consumer protection organization. In E.-J. Abbots, A. Lavis and L. Attala (eds) *Careful Eating: Bodies, Food and Care*. Farnham: Ashgate.

foodwatch (2016) About foodwatch. Available at: www.foodwatch.org/en/about-foodwatch/ (accessed 8 August 2016).

Fox, N. J. (2015) Personal health technologies, micropolitics and resistance: A new materialist analysis. *Health* 1363459315590248.

Gabriel, Y. and Lang, T. (2015) *The Unmanageable Consumer*. London: Sage.

Goodman, D. and Goodman, M. K. (2009) Food networks, alternative. In R. Kitchin and N. Thrift (eds) *International Encyclopedia of Human Geography*. Amsterdam: Elsevier, pp. 208–220.

Langlois, G., Elmer, G., McKelvey, F. and Devereaux, Z. (2009) Networked publics: The double articulation of code and politics on Facebook. *Canadian Journal of Communication*, 34(3): 415–434.

Law, J. and Singleton, V. (2013) ANT and politics: Working in and on the world. *Qualitative Sociology*, 36(4): 485–502.

Lekakis, E. J. (2014) ICTs and ethical consumption: The political and market futures of fair trade. *Futures*, 62: 164–172.

Lockie, S. (2006) Networks of agri-environmental action: Temporality, spatiality and identity in agricultural environments. *Sociologia Ruralis*, 46(1): 22–39.

Morris, C. and Kirwan, J. (2011) Ecological embeddedness: An interrogation and refinement of the concept within the context of alternative food networks in the UK. *Journal of Rural Studies*, 27(3): 322–330.

O'Connor, C. (2013) New app lets you boycott Koch Brothers, Monsanto and more by scanning your shopping cart. *Forbes*, 14 May 2013. Available at: www.forbes.com/sites/clareoconnor/2013/05/14/new-app-lets-you-boycott-koch-brothers-monsanto-and-more-by-scanning-your-shopping-cart/#3263deda2c82 (accessed 1 July 2017).

Roe, E. J. (2006) Things becoming food and the embodied, material practices of an organic food consumer. *Sociologia Ruralis*, 46(2): 104–121.

Ruppert, E., Law, J. and Savage, M. (2013) Reassembling social science methods: The challenge of digital devices. *Theory, Culture & Society*, 30(4): 22–46.

Sayes, E. (2014) Actor–Network Theory and methodology: Just what does it mean to say that nonhumans have agency? *Social Studies of Science*, 44(1): 134–149.

Schneider, T., Eli, K., McLennan, A., Dolan, C., Lezaun, J. and Ulijaszek, S. (forthcoming) Governance by campaign: The co-constitution of food issues, publics and expertise through new information and communication technologies. *Information, Communication and Society.*

Siniscalchi, V. and Counihan, C. (2014) Ethnography of food activism. In C. Counihan and V. Siniscalchi (eds) *Food Activism: Agency, Democracy and Economy.* London: Bloomsbury, pp. 3–12.

Van Dijck, J. and Poell, T. (2013) Understanding social media logic. *Media and Communication,* 1(1); 2–14.

Vegh, S. (2003) Classifying forms of online activism. In M. McCaughey and M. D. Ayers (eds) *Cyberactivism: Online Activism in Theory and Practice.* London: Routledge, pp. 71–95.

Afterword

The public's two bodies – food activism in digital media

Javier Lezaun

In December 1997, the United States Department of Agriculture (USDA) launched an initiative in what it called 'electronic government'. After years of discussions with experts on the proper definition of the term 'organic food', the agency decided to publish on the Internet its proposed set of standards for the certification of these products, inviting the public to offer its views on the draft legislation. The proposal was the first time the US government had attempted to define the category of 'organic'. Thanks to the innovative use of the Internet, this was also presented as 'the first fully electronic rule-making for a major regulation in federal history' (Friel, 1998).

The public reaction was, by all accounts, overwhelming. During the consultation period, the USDA received more than 275,000 comments. Secretary of Agriculture Dan Glickman noted that this response was '20 times greater than to anything ever before proposed by the Department'. The vast majority of the comments were highly critical of the proposal, particularly of the USDA's plan to allow under the label 'organic' several controversial products and practices (genetically modified organisms, food irradiation, or the use of 'municipal biosolids' – sewage sludge – as fertilizer). Following the storm of negative feedback, Secretary Glickman withdrew the proposal and promised to develop a new set of standards that would better reflect the sentiment expressed by the public.

At the time this was heralded as a landmark event in the rise of 'digital democracy'. Much was made of the total number of comments submitted – news reports noted that the agency had received more than one comment per minute during the four-month consultation period. Yet the Internet played only a marginal role in the scale and intensity of the public response. Less than 10 per cent of comments were submitted via the USDA's website (Shulman, 2003); the vast majority were received via postal mail or fax, and were the result of energetic letter-writing campaigns coordinated by consumer organizations and organic producers, an 'old-fashioned' form of activism, aided in this case by extensive coverage of the issue in newspapers and other traditional media. The USDA's about-face said a great deal about the growing power of what Michael Pollan (2001) would call the 'organic-industrial complex', but very little about the potential of the Internet as a new medium of food activism.

Still, the process set up by the USDA included a small feature that served as a harbinger of things to come. The agency scanned all the comments it received and entered them into a searchable online database while the consultation period was still open, thus allowing members of the public to read other citizens' submissions and write their own comments in response. In this way, a channel created to facilitate unidirectional communication from citizens to government began to display its potential as a platform on which individual views could converge, influence one another, and be aggregated as a new sort of public.

The case of the USDA's organic standards is also exemplary for the manner in which the creation of a new digital space for advocacy went hand in hand with the careful framing of what that advocacy was expected to be about. The USDA justified its interest in eliciting public reactions to its draft legislation – and the use of the Internet to do so – with the argument that the proposed rules were exclusively concerned with the appropriate *marketing* of organic products, and that this question was strictly separate from any discussion of those products' health or environmental impact. The latter were a matter for scientific experts, the USDA argued, while the former allowed, indeed demanded, significant public consultation, since the standards would by definition fail if the public found them too divergent from their own definition of 'organic'.

Thus, the very rationale that the USDA used to justify its experiment in digital public consultation defined the terms of the debate in ways that marginalized key tenets of the social movement for organic agriculture. Namely, that the quality of 'organic' spoke not only to a set of agricultural inputs or a specific process of food production (and to how to advertise those inputs and that process to consumers), but also to the nutritional, social and environmental benefits that obtained from producing foods in an ecologically sustainable way. Narrow technical standards of production were amenable to a process of public consultation that privileged quantitative metrics of consent (or opposition). More expansive considerations about the meaning of 'organic' within agro-ethical visions of social transformation overflowed the USDA's grid of responses and were not computed in the agency's tabulation of public opinion (Buttel, 1997; Goodman, 2000; Guthman, 2014; Vos, 2000).

When the USDA issued a revised legislative proposal in 2000, the most controversial aspects of the original draft were gone. (In particular, the use of genetically modified organisms was deemed to be incompatible with the certification of a food product as 'organic'.) What remained was the emphasis on organic certification as a *marketing* tool, and the careful demarcation of this issue from matters of food safety, nutritional quality, or environmental stewardship. After describing the changes introduced in the new proposal as 'a living example of our democracy at work', Secretary Glickman emphasized once again the restricted nature of the consultation exercise: 'The organic label is a marketing tool. It is something that I think consumers want. It is not a statement by the government about food safety. Nor is "organic" a value judgment by the government about nutrition or quality' (Glickman, 2000).

This now seemingly archaic moment in the history of digital food activism – launched before the majority of U.S. citizens had regular access to the Internet – displays some of the tensions that run through this volume. To use the distinction introduced by the editors, this was an early example of 'Internet-enabled' advocacy rather than 'Internet-based' collective action (Chapter 1 in this volume; Chapter 11 in this volume). Yet it offered a glimpse into how the specific affordances of digital media would transform activism over the following two decades, and signalled the importance of the struggle to define the limits – discursive and material – of these new forms of advocacy.

If we take seriously, as this book does, the hypothesis that the digital realm offers a distinct medium of activism, then we are forced to accept that this medium will come with its own forms of freedom and constraint, openness and closure, transparency and opacity. This is a medium in which widespread dissemination of information is compatible with ever-growing concentrations of power, and where the capacity to tailor messages to highly specific micro-audiences is both an instrument of political mobilization and the basis for unprecedented levels of marketing precision. Any attempt to analyse digital food activism must thus start with this ambivalence – or rather multivalence – of the digital. The most creative examples of food activism are those that use this ambivalence to their own advantage.

Analytics of participation

Digital food activism can be seen within a set of broader transformations in the governance of food systems, what Xaq Frohlich (2017) calls 'the informational turn in food politics'. This turn pre-dates the Internet, and is founded on the idea that food markets are best governed by framing the informational exchanges between producers and consumers, rather than by imposing on them a command-and-control regulatory architecture. The figure at the centre of this regime, at least nominally, is the 'active', 'empowered', 'responsible' or 'reflexive' consumer, able to weigh the multiple choices available to him/her in the marketplace with the help of detailed information about individual food products. Changes in labelling rules, and in practices of food categorization more generally, represent the key mechanism for the improvement of diets and the transformation of production systems (Schleifer, 2013). Struggles about the material composition of foods become then indistinguishable from struggles about the *informational qualification* of those foods (cf. Callon *et al.*, 2002; see also Lezaun and Schneider, 2012).

Digital activism makes the informational ecology of food systems increasingly complex, dynamic, and unruly. Operating in a crowded digital public sphere, activists must compete with actors that have a direct economic stake in the qualification of food products and sophisticated communication strategies of their own. In fact, to the extent that control over information becomes key to the design of food markets, the boundary between activism and advertising, food advocacy and food provision becomes blurred. The small coffee producers in

Oaxaca that Sarah Lyon describes Chapter 4 in this volume, actively seek to straddle these distinctions. Their power to constitute themselves as 'producers' depends on their ability to establish new communication links with affluent consumers, including digitally enabled forms of traceability and political messaging. In this regard, digital food activism can be much more than a tool to empower consumers or a means of politicizing their consumption choices. It can effectively serve to shift roles within the food system – from wholesaler to retailer, from consumer to producer (or prosumer), from regulated to regulator. At stake, then, is not simply the qualification of specific market products, but the very architecture of those markets.

As with any other form of advocacy, however, food activism is transformed when it is embedded in digital media, as the conditions of its own visibility change. For one, access to the information that activists produce is filtered through software features, search functions, and algorithmic designs – or, in some cases, directly censored by the states and corporations that control the physical infrastructures of the digital realm. As a result, activists become 'content providers' in a communication system where the ability to connect with relevant audiences and constituencies is always at the mercy of powerful incumbent actors. The attention of those audiences and constituencies represents the most valuable resource, but it is a scarce and fragile one, always mediated by the interests of invisible third parties (Tufekci, 2017).

As it moves onto digital platforms, moreover, activism becomes entangled in what we might call the analytics of participation. Participation through digital media is eminently trackable and measurable, to an extent that traditional forms of advocacy and collective action never were. Any instance and any modality of activism leave a trace that can be identified, aggregated, and analysed; it produces a 'footprint' of involvement with an issue that those with the required analytical capacities can use to produce a profile of the individual in question or to draw points of connection between him/her and other citizens. Whether it leads to social change or not, then, digital participation always produces data, and the capacity to access and use that data becomes increasingly central to activist and counter-activist strategies. An organization like the German non-governmental organization foodwatch (Chapter 11 in this volume; see also Schneider *et al.*, forthcoming), which operates primarily through digital media, will have access to detailed information about its audiences – their number and geographic location, the distribution of interest across the range of campaign issues the organization addresses, or the extent and speed with which members of the public access and share a specific item of advocacy. The medium of communication then comes to shape not only the specific tactics used by the organization to deliver its messages, but the strategic processes through which it crafts those message and imagines its audiences.

Access to this data, however, and the capacity to analyse it, are distributed in radically unequal terms. It is safe to assume that the digital platforms through which an app like Buycott are made available will have better and more granular information than its developers about its users and their habits (Chapter 11 in this volume; see also Eli *et al.*, 2016). In fact, they will possibly have greater insight into how, where, and when the app is used than the users themselves. So while the developer of the app might position the tool as a 'neutral space', open to multiple activist agendas, the platforms that distribute it and track its use are highly interested actors driven primarily by commercial purposes. Critically, they are in a position to identify behavioural patterns and connections between users that are fundamentally opaque both to the individual citizen, and to the campaign organization that seeks their mobilization (see also Chapter 2 in this volume).

In other words, while not every gathering on a digital media platform will come to constitute an 'expressive collectivity' (Couldry, 2015), the aggregation of individual acts of advocacy will increasingly represent a meaningful population for data-analytical – and commercial – purposes. Thus the 'logic of connective action' identified by Bennett and Segerberg (2013) in digital media presents two different manifestations: it refers, on the one hand, to the ability of individuals to produce personalized content and establish links with like-minded citizens; but it also describes, on the other hand, the analytic capacity of organizations and institutions to aggregate data and draw patterns of connectivity (actual and potential) within large sets of users. In some cases, these two logics may intersect – in Chapter 5 in this volume, Katharina Witterhold) mentions the example of a consumer who uses her loyalty card to pay for vegan products in the hope that these purchases will be tracked and thereby signal to marketeers the growing demand for these products. Yet more often than not, the two logics will diverge, as activist groups in positions of structural weakness confront powerful incumbents motivated by radically different goals and armed with forbidding amounts of financial and technical resources, including proprietary data-analytical capacities. Hence the emphasis of many of the best-known examples of digital food activism on the sort of 'disruptive practices' Melissa Caldwell describes in Chapter 2 in this volume, or what Eva Giraud calls 'tactical interventions' in Chapter 7: attempts to interfere in complex communication ecologies by modulating the affordances of particular media, a sort of digital weapon of the weak intended to counteract the growing power differentials in this realm.

The public's two bodies

Even if the digital public sphere is increasingly structured by powerful commercial interests, that does not make it a fully controlled or tame social space. As Noortje Marres (2017) has argued, a distinct feature of digital networked content is its dynamic quality, the manner in which its peculiar form of sociality tends to exceed the prescriptive power of those attempting to manage it. The same trait that generates the analytical richness of digital content, its cheap and quick shareability, also ensures its political potency. Recent anxieties about the proliferation of 'fake news', or the emergence of virulent ideological subcultures specific to social media formats (Nagle, 2017), are symptoms of the unruliness of the digital realm, but also of its potential to create new and unexpected political configurations.

In exploring this political potential, we can extend to digital food activism the notion of the 'recursive public', which Chris Kelty (2008) developed to account for the cultural significance of the Free Software movement. 'Recursive publics,' Kelty writes, 'are publics concerned with the ability to build, control, modify, and maintain the infrastructure that allows them to come into being in the first place' (ibid.: 7). Kelty's concept describes collectives that attend to 'the radical technological modifiability of their own terms of existence' (ibid.: 3), and that use that awareness of malleability to explore new forms of community (see also Dunbar-Hester, 2014).

Recursive forms of activism, in this sense, are not concerned, at least not exclusively, with any particular *object* of contention, but rather with the ebbs and flows of their own *capacity* to intervene in the infrastructures that define their public character (cf. Warner, 1990). Digital food activism thus raises the possibility of a *doubly-recursive* public, a public that attends both to the conditions under which the food it consumes are produced, and to the systems that generate and disseminate the information that underpins its food choices. This is a public, in other words, concerned with its own *two* bodies, the material-metabolic and the informational-discursive.

The possibility of such a form of collective action can be glimpsed in several of the chapters in this volume, and is evident in several strands of contemporary food activism. Loosely organized, Internet-based collectives like the Label-It-Yourself (LIY) movement use the official communication channels of the food system to disseminate unofficial, counter-commercial messages (Lezaun, 2014). By sticking their own labels on the packages of food products on supermarket shelves, LIY activists introduce a bit of disruptive noise in the carefully policed interface of producers and consumers. Label designs – and instructions for how to re-label products furtively in closely watched spaces like that of the supermarket – are freely available on the Internet, making this format of advocacy available for any particular protest issue, whether food-related or not.

A similar use of digital media as a point of convergence for diverse activist strategies characterizes the food producers gathered in the Farm Hack collective, an online community dedicated to developing open-source agricultural tools and equipment. Participants in Farm Hack – farmers, but also engineers, designers, or programmers – make their designs freely available through online repositories and offline meet-ups, and license them under both attribution and share-alike creative commons provisions, ensuring that any derivative work is released into the public domain without any additional restrictions. The close relationship that Farm Hack establishes between the principles that animate its design practice (e.g. replicability, affordability, design for disassembly), and those that guide its community-building efforts (e.g. horizontal, user-driven decision-making; share-alike licensing) speaks to the recursive nature of this example of what Michael Carolan (2016) calls agro-digital governance.

Initiatives like Label-It-Yourself or Farm Hack intervene at the critical, informational nexus of food systems, and offer an explicit counterpoint to the centrality of proprietary technology and private data to their organization (Bronson and

Knezevic, 2016). In this sense, they are heirs to an old and venerable tradition of agro-food advocacy: the long history of social movements campaigning for the free circulation of seeds, movements that saw in the communal sharing of this crucial resource – *both* physical commodity *and* informational node – the most direct way of enacting a radical alternative to corporately controlled, industrialized food production (Aoki, 2009; Kloppenburg, 2005).

The open source ethos of initiatives, like Label-It-Yourself, Farm Hack, and other initiatives discussed in this volume, implies a recognition of the potential of digital platforms to create new forms of community. The focus of activism is no longer – or at least not primarily – the qualification of specific markets products, or the fashioning of a different kind of consumer, but the production of a new social identity: an *activist* identity whose ethical horizon is not contained by existing market architectures, but spills over into the creation of new polities, however fragile or transient these might be.

References

Aoki, K. (2009) 'Free seeds, not free beer': Participatory plant breeding, open source seeds, and acknowledging user innovation in agriculture. *Fordham Law Review*, 77(5): 2275.

Bennett, W. L. and Segerberg, A. (2013) *The Logic of Connective Action: Digital Media and the Personalization of Contentious Politics*. Cambridge: Cambridge University Press.

Bronson, K. and Knezevic, I. (2016) Big data in food and agriculture. *Big Data & Society*, 3(1), first published online: 20 June 2016, DOI: 10.1177/2053951716648174.

Buttel, F. H. (1997) Some observations on agro-food change and the future of agricultural sustainability movements. In D. Goodman and M. Watts (eds) *Globalising Food: Agrarian Questions and Global Restructuring*. Hove: Psychology Press.

Callon, M., Méadel, C. and Rabeharisoa, V. (2002) The economy of qualities. *Economy and Society*, 31(2): 194–217.

Carolan, M. (2016) Agro-digital governance and life itself: Food politics at the intersection of code and affect. *Sociologia Ruralis*. First published: 17 November 2016. DOI: 10.1111/soru.12153.

Couldry, N. (2015) The myth of 'us': Digital networks, political change and the production of collectivity. *Information, Communication & Society*, 18(6): 608–626.

Dunbar-Hester, C. (2014) *Low Power to the People: Pirates, Protest, and Politics in FM Radio Activism*. Cambridge, MA: MIT Press.

Eli, K., Dolan, C., Schneider, T., and Ulijaszek, S. (2016) Mobile activism, material imaginings, and the ethics of the edible: Framing political engagement through the Buycott app. *Geoforum*, 74: 63–73.

Friel B. (1998) Real world results. *Government Executive*. 1 December 1998. Available at: www.govexec.com/magazine/1998/12/real-world-results/7618/ (accessed 13 October 2017).

Frohlich, X. (2017) The informational turn in food politics: The US FDA's nutrition label as information infrastructure. *Social Studies of Science*, 47(2): 145–171.

Gillespie, T. (2014) The relevance of algorithms. In T. Gillespie, P. J. Boczkowski, and K. A. Foot (eds) *Media Technologies: Essays on Communication, Materiality, and Society*. Cambridge, MA: MIT Press, pp. 167–193.

Glickman, D. (2000) USDA Secretary Dan Glickman comments at release of national organic standards. 20 December 2000. Available at: www.nationalorganiccoalition. org/_literature_137704/CFS (accessed 17 October 2017).

Goodman, D. (2000) Organic and conventional agriculture: Materializing discourse and agro-ecological managerialism. *Agriculture and Human Values*, 17(3): 215–219.

Guthman, J. (2014) *Agrarian Dreams: The Paradox of Organic Farming in California.* Berkeley, CA: University of California Press.

Kelty, C. M. (2008) *Two Bits: The Cultural Significance of Free Software.* Durham, NC: Duke University Press.

Kloppenburg, J. R. (2005) *First the Seed: The Political Economy of Plant Biotechnology.* Madison, WI: University of Wisconsin Press.

Lezaun, J. (2014) Iconoclasm in the supermarket. *Limn 4: Food Infrastructures.* Available at: http://limn.it/issue/04/ (accessed 13 October 2017).

Lezaun, J., and Schneider, T. (2012) Endless qualifications, restless consumption: The governance of novel foods in Europe. *Science as Culture*, 21(3): 365–391.

Marres, N. (2017) *Digital Sociology: The Reinvention of Social Research.* Cambridge: Polity Press.

Nagle, A. (2017) *Kill All Normies: Online Culture Wars from 4Chan and Tumblr to Trump and the Alt-Right.* London: Zero Books.

Pollan, M. (2001) Behind the organic-industrial complex. *New York Times.* 13 May.

Schleifer, D. (2013) Categories count: Trans fat labeling as a technique of corporate governance. *Social Studies of Science*, 43(1): 54–77.

Schneider, T., Eli, K., McLennan, A., Dolan, C., Lezaun, J. and Ulijaszek, S. (2017) Governance by campaign: the co-constitution of food issues, publics and expertise through new information and communication technologies. *Information, Communication & Society*, 1–21, Published online: 23 August 2017, Available at: http://dx.doi.org/10.108 0/1369118X.2017.1363264

Shulman, S. W. (2003) An experiment in digital government at the United States National Organic Program. *Agriculture and Human Values*, 20(3): 253–265.

Tufekci, Z. (2017) Twitter and Tear Gas: The Power and Fragility of Networked Protest. New Haven, CT: Yale University Press.

Vos, T. (2000) Visions of the middle landscape: Organic farming and the politics of nature. *Agriculture and Human Values*, 17(3): 245–256.

Warner, M. (1990) *The Letters of the Republic: Publication and the Public Sphere in Eighteenth-Century America.* Cambridge, MA: Harvard University Press.

Index

Page numbers in *italics* denote tables, those in **bold** denote figures.